Mushrooms of the upper Midwest

A Simple Guide to Common Mushrooms, Growing Gourmet and Medicinal Mushrooms, Mycophilia,

By
ANDREW PAUL

LEGAL DISCLAIMER

© Copyright 2019 by Andrew Paul - All rights reserved

This document is geared towards providing exact and reliable information in regards to the topic and issue covered. The publication is sold with the idea that the publisher is not required to render accounting, officially permitted, or otherwise, qualified services. If advice is necessary, legal or professional, a practiced individual in the profession should be ordered.

In no way is it legal to reproduce, duplicate, or transmit any part of this book in either electronic means or in printed format. Recording of this publication is strictly prohibited and any storage of this document is not allowed unless with written permission from the publisher. All rights reserved.

The information provided herein is stated to be truthful and consistent, in that any liability, in terms of inattention or otherwise, by any usage or abuse of any policies, processes, or directions contained within is the solitary and utter responsibility of the recipient reader.

Under no circumstances will any legal responsibility or blame be held against the publisher for any reparation, damages, or monetary loss due to the information herein, either directly or indirectly.

The author owns all copyrights not held by the publisher.

The contents of this book are for informational purposes solely. The presentation of the information is without contract or any type of guarantee assurance. Any trademarks and brands within this book are for clarifying purposes only and are owned by the owners themselves and not affiliated with this document.

Table of Contents

Introduction 1

What Exactly Is A Mushroom? 3

Top 5 Easy-To-ID Edible Mushrooms
for Beginners 5

The Mushroom At The End Of The World 12

Magic Mushrooms 23

The Type Of Magic Mushroom You
Should Consume 28

Everyday Experiences Of Magic
Mushroom Trips 30

How To Find Mushrooms 41

A Simple Guide To Common Mushrooms 51

Delve Deeper Into The History Of Mushroom 54

The Grower's Guide To Psilocybin Mushroom 59

Healing Mushrooms 74

Mushrooms For Cancer 81

    Ganoderma Mushroom 92

Acupuncture & Mushroom Nutrition -
Secret To Longevity 103

Grow Mushrooms For Food And Other Reasons 110

Learning How to Farm Mushrooms 123

How To Grow Mushrooms - Learn About Growing Mushrooms	133
Types Of Mushrooms	137
The Mighty Mushroom	149
Mushroom Varieties And Their Uses	167
Conclusion	172

Introduction

There's something otherworldly and mysterious about the world of fungi, which is part of the reason diving wild mushroom hunting can be so intimidating. Even after becoming an avid plant forager myself, I held off from expanding into mycophagy, or wild mushroom foraging, assuming it to be more "advanced" than plant foraging, and more likely to result in mistakes.

However, in this book, we will explore the Secret of Mushrooms of the Upper Midwest. Stories of toxic mushrooms, psychedelic delirium, and names like made me feel like I was better off just sticking with wild plants.

How silly I was! The truth is, while there are mushrooms with toxic lookalikes that are hard to differentiate, the same basic foraging rules apply to mushrooms that use to plants. Become super-familiar with all the different identification characteristics; don't eat something unless you have a 100% identification.

Start with easy ones before you try to identify more delicate species. Start by only eating a small amount the first time. Other details, like not carrying different mushroom species in the same bag, are quickly learned. A careful, toe-in-the- water-first approach to mycophagy reveals that mushroom foraging is inherently no more dangerous than plant foraging.

Now I feel silly for my participation in modern culture's irrational fear of eating delicious, nutritious wild mushrooms. Knowing just a few varieties massively expands your chances in a survival situation. But it also makes you less dependent on the industrial food system, which uses massive quantities of fuel and water to grow and transport nutritionally-inferior produce to supermarket shelves.

Finding any mushroom is a thrill, as every mushroom your find is an extraordinary gift from Mother Nature. You have to be in just the right place, at only the right time, under just the right conditions.

Let's get started!

What Exactly Is A Mushroom?

Mushrooms are a type of fungus. There are many different kinds of fungi, including molds and crusts, as well as more developed types that have a stalk and a cap. Mushrooms are distinct from plants because they do not possess chlorophyll, the green pigment that allows plants to manufacture sugar from the sun's energy; they need to absorb their food from the environment in which they live.

A mushroom is the sexual organ of a much larger organism called fungi. The primary purpose of the mushroom is to produce spores for sexual reproduction. As a mushroom grower, I like to consider the mushroom as fruit, and you will often see it referenced as a fruiting body. This is because the truffle has more functions than just sexual reproduction. The mushroom is the fruit of a much larger organism called fungi.

Fungi use fibers called hyphae (that as a group are referred to as mycelium), to take in food. The mycelium can remain dormant under the ground for many reasons, similar to the roots of plants. Each hypha that is sent out makes its way through earth/wood/plant matter until it reaches the surface.

During the organism's specific growing season, the hyphae develop into mature structures capable of reproducing spores. The structure that you usually see above the ground is the part of the mushroom that is producing and dispersing spores.

Each spore is a single cell that is capable of sending out a hypha that will develop into a group and form its mycelium. If the hypha of one spore meets up with the hypha of another, it begins the sexual process of spore production through individual spore-producing cells.

Mushrooms sprout suddenly and proliferate; many will be rotten or inedible just hours or days after you find them. The window for a successful harvest is open for only a tantalizingly brief period. Meanwhile, you compete with millions of hungry insects for your nibble of the delicious prize, making a successful mushroom forage even more rewarding.

Another fascinating aspect of mycophagy is the potential heft of your harvest – with many species. It's familiar to net five, ten, or even twenty pounds of mushrooms. When you're foraging plants, it's nearly impossible to get that kind of weight in a single swoop.

To help you get started, below are my picks for the five easiest-to-identify edible mushroom species. In addition to being beginner-friendly, each species on my list comes in potentially enormous harvests of ten pounds or more!

Top 5 Easy-To-ID Edible Mushrooms for Beginners

➢ Morels

One of the most prized edible mushrooms in the world, morels can fetch upwards of $40 per pound on the open market. They're also incredibly easy to identify!

Look out for the signature deep pits and sharp ridges of the morel's yellowish to black or brown caps.

The superficially-similar "false morel" has only ridges or waves, rather than the bottomless pits of true morels. Another way to distinguish true morels from false morels is to cut one open and make sure the mushroom is hollow and from the tip of the cap through the stem. True morel caps also tend to be longer and narrower, while false morels often have more full, stubbier, chunkier caps.

Morels have a vibrant and meaty taste. They appear in late spring through early summer and like to grow on south-facing hills on the edges of woods. They also pop up around dead ash, elm, and other trees. Like many species, remember where you found them and check at the same time next year, as they often return each season.

Just be careful who you tell about your discovery or the word may spread, and there will be none left for you — according to Alexander Schwab, author of the excellent book Mushrooming With Confidence, morels "are so prized that details of the sites where they are found are left in wills or passed on from the deathbed."

➢ Chicken of the Woods

Also known as the sulfur shelf mushroom, "Chicken of the Woods" is named for its mild, versatile flavor (like chicken), and is easily identified by its dramatic yellow and orange shelves. From Brooklyn to India, nothing else on the trail quite sticks out like a bright, fire-colored Chicken of the Woods hanging off the side of a dead tree or stump.

Chicken of the Woods can fruit on a variety of trees, but most commonly grows on oaks. Autumn is its favorite season in most areas. Its surface is rubbery and often flesh juicy, but becomes woodier and more brittle as it ages. The shelves also become flatter, thinner, and less brightly-colored in older specimens.

The smooth, yellow undersides of each cap contain many tiny pores, or tubes, numbering about three pores per millimeter of surface area. Occasionally the Chicken mushroom can grow in rosettes instead of distinct shelves.

The younger it is when you harvest it, the more tender and delicious it will be. A similar and also-edible species have nearly identical characteristics, but is paler, with a white pore surface instead of the distinctive orange. Other than that, the only potential lookalike is Berkeley's Polypore, which is also edible.

Many chicken mushroom hauls can have you walking away with ten or twenty pounds of mushrooms, while still leaving some behind for other critters.

➢ Oyster Mushrooms

A common sight both in grocery store produce sections, and on trees, the oyster mushroom is a delicious, nutritious, and easy-to-identify wild treat. It can grow from winter through summer – it all depends on the oyster sub-species, tree type, and foraging region.

Look for overlapping, tiered oyster or fan-shaped caps growing on deciduous trees and stumps. Caps can be colored anywhere from creamy off-white to gray or brown. The mushroom's structure is always a tiered arrangement of fan- shaped caps, and the gills are still creamy off-white to brownish. Caps are smooth, often moist, and sometimes have wavy edges. Younger specimens have a more rounded cap shape.

Stems are short and tend to fuse into a single stalk, connecting several of the caps into one stem. The gills will run nearly down the stem. Make sure you see no sign of a "veil," which is a skirt-like covering between the stem and the edges of the cap.

By the time gills are brown, the oysters are too old to consume, as the texture will be dry and unappealing, so look for the fan-shaped caps and lighter-colored gills while hunting.

➢ Hen Of The Woods Mushroom

Hen of the Woods is often discovered when hunters recognize its large, distinctive brownish clumps at the bases of oak trees and stumps. This prized edible mushroom consists of overlapping fan-shaped grayish to brownish caps that, when a specimen is cut in half, are revealed to descend into a single, thick base stem. Interior flesh is whitish, sometimes with a tinge of yellow.

Inspect the bottom of the caps, and you'll see a network of spore tubes. Hen of the Woods doesn't have poisonous lookalikes. Superficially-Similar looking species like the Cauliflower Mushroom and Umbrella Polypore are delicious edibles as well.

➢ Puffball Mushroom

Puffballs, a scourge for lawn owners everywhere, are a mushroom hunter's treasure. They have a delightful, marshmallow-like texture, and large puffball species can quickly become as large as volleyball. puffballs-emitting-spores

Some are smooth, some are spiny, some are warty, and some develop crusty mosaic patterns on their skin as they age, but a few common characteristics can be used to ensure your puffball is of the edible variety.

First, the inner flesh should be uniformly white and marshmallow-firm. The exterior skin can be slightly off-white. If there is a discernible stem, the internal tissue should continue consistently throughout. In other words, puffballs don't have a right, separate "stem" structure just a narrower base section with nothing separating it from the top part of the mushroom.

Puffballs never have a strong smell. Beware of baby amanitas, which look similar but lack the true puffball's pure, completely uniform white inner flesh when sliced in half. As puffballs age, they darken and release dramatic clouds when poked or kicked. By this point, they are far beyond the point where they can be eaten.

Playing with aging puffballs can help spread the spores, be careful not to breathe them in!

Basics of Cooking And Storing Mushrooms

For most wild mushrooms, sautéing in butter with salt and pepper is a great way to test the flavor for the first time. Dehydrating and grinding them up is usually useful for starting a creamy soup and cooking stock that will have a long shelf life. Giant Puffballs are a bit more unique-tasting, but make for excellent "Puffball Parmesan."

Chicken of the Woods, you can slice up, season, and use in stir-fries, tacos, and anything else that'd use chicken strips. The earthy flavor of Hen of the Woods, known as "Maitake" in Japan, has lent itself well to many Japanese dishes.

Morels are fantastic with eggs or steak, or chopped and tossed into vegetables. Just use caution if you're a drinker, as some people report stomach upset when combining come morels, especially black morels, with alcohol. Oyster mushrooms are versatile. An easy way to figure out what to do with most wild mushrooms is to look at recipes that call for store-bought varieties. Chances are, using wild mushrooms in the same methods will take those dishes to the next level in terms of both tastiness and nutrition.

Some varieties freeze better than others, and some rot exceptionally quickly. Transport and store in paper bags only, and for the best results possible, you can't go wrong cooking and eating them the same day you find them! Mushrooms

Piqued Your Interest?

Here are a few excellent resources worth checking out:

Edible Wild Mushrooms of North America: This field guide is older but among the most comprehensive. It covers the selection, processing, and cooking (A-Z). It's indispensable and an excellent resource to have on hand.

For region-specific guides, check out:

Mushrooms of the Upper Midwest: Well illustrated and easy to navigate if you are in the Upper Midwest (Michigan, Illinois, Ohio, Wisconsin, etc). Field Guide to Wild Mushrooms of Pennsylvania and the Mid-Atlantic: Another handy resource if you correctly are operating in this area. A Field Guide to Edible Mushrooms of the Pacific Northwest: The Pacific Northwest has some of the most diverse (and region-specific) arrays of mushrooms. If you live in the area, definitely an excellent addition.

The Mushroom At The End of The World

In a world that is falling apart (no further elaboration needed), how shall we understand the dynamics of survival and collaboration? How does life persist and flourish in a world that is hellbent on commodifying and privatizing every aspect of human relations and the natural world?

Anna Lowenhaupt Tsing's book The Mushroom at the End of the World is subtitled "On the possibility of life in capitalist ruins." She's not talking about the future when the industrial economy has completely collapsed. She's talking about the present, in places where collapse is already well underway.

The book describes how alternative economies not the neat, tidy 'sharing' economies we idealists like to write about, but the underground economies that emerge out of necessity, and always have — evolve, and the lessons they have for us as our civilization culture continues to crumble.

It's a radical and equanimous book — it leaves judgment about economics and justice and fairness to others, and instead describes what is, in well-researched, gritty detail. To do so Anna introduces many new terms to the lexicon of economics, necessary she says because analytic, mechanical models of economics fail to describe how economies really work.

So, a whole new vocabulary of holistic terms is needed, words that describe the inseparable interdependence between creatures and environments, instead of the economist's usual oversimplified model of 'resources' as something separate, and mechanisms of purported control.

To show this interdependence, she describes how one tiny piece of the economy works — the harvesting and distribution of matsutake mushrooms. She then shows how staggeringly complex and uncontrollable the workings of this completely self-evolved and self-organized economy are, and in so doing demonstrates that every aspect of our economy works like this — the belief that we can fully understand and 'manage' an economy at any level is shown to be complete hubris.

A small and incomplete but core part of this economy is illustrated in the diagram a bove. Here are its essential components:

Like almost everything in our modern culture, the creation of wealth through matsutake mushrooms starts with a salvage operation. Salvage is the process of exploiting value outside the capitalist supply chain. Drilling for oil is salvage.

Harvesting crops is salvage. Contracted labor is salvage.

The capitalist economy has outsourced almost all salvage operations to independent contractors to minimize risk. Capitalists attempt to control these contractors by controlling the elements of the supply chain as it enters and leaves the industrial economy (yellow area on chart).

So, Chinese slave labor garment factories are bonded to multination "brand" corporations through license agreements. They don't care what goes on in the grey "salvage economy" area — let the salvagers argue with the environmentalists, lawyers, regulators, and social justice advocates.

Their job is to accumulate the salvage, alienate it (make it unrecognizable as to source, as much as possible, so customers are in the dark as to what factory farming or slavery or genetic alteration or other operation was involved — hence their opposition to labelling initiatives), and commoditize it for the market — the part of the supply chain that is low-risk and under their control.

What we read about in economics books is just what happens in the yellow area of the chart, the most insignificant part, but it's the part where almost all profit accumulates (as required by corporate charter), and the most expensive part (where the vast oligopolistic price increases are in the supply chain).

Much of the book describes the life and process of the matsutake pickers in the forests of Eastern Oregon, to convey how unfathomably complex the essential salvage economy is. Forestry policies over the years (themselves dependent on innumerable variables) have resulted in ruined forest wasteland in much of Eastern Oregon.

Timber companies clear-cut much of the state then attempted to replant monoculture tree species to keep the industry going, and eventually, due to many other factors, abandoned the effort, leaving a mess. First Nations had paid close attention to the ecology and had found sustainable ways (including controlled burning) to manage the forests (as have indigenous peasant groups in most countries with forests). But when the First Nations peoples were slaughtered and driven out, their skills were lost.

The unnatural mess left behind by the timber companies creates huge fire dangers that modern fire prevention measures exacerbate. But it would take a century (which our civilization doesn't have) to manage these forests back to health through knowledgable 'disturbances' (the term Anna uses for interventions, planned and unintentional).

As with our modern industrial, agricultural policies (what is accurately called 'disaster agriculture' since it uses wholesale flooding, burning, poisons and plowing to keep unnatural monocultures going), we have no idea how to 'disturb' ecosystems in healthy, sustainable ways, so we just keep creating monoculture messes and abandoning them when yields disappear.

Creating such disturbances, carefully and modestly, is the essence, Anna says, of intervening in an ecosystem for healthy succession over decades — real permaculture. The ruined mess in Eastern Oregon has allowed "uneconomic" fir and pine trees to flourish, and matsutake mushrooms thrive in such forests.

At the same time, Japanese forestry policies and programs (which are in turn a result of western occupation after WW2 and other complex factors) have largely eliminated firs and pines there in favor of more profitable species, accidentally killing off the matsutake, a centuries-old staple, and delicacy in Japan, in the process. So suddenly, with no local supply and new supply in Oregon, a new economy was born in the ruined forests of Oregon, and also in the peasant forests in China and elsewhere.

Except for a few war vets, whites are not inclined to pick mushrooms in the forest for a living. But Southeast Asians, many of them refugees to the western US from the Vietnam war and other American wars, are skilled at picking them, and delighted for an alternative to the discrimination and soulless labor in American cities, so they flocked to the freedom of the Oregon forests, creating entire communities along the ethnic lines (Lao, Khmer, etc.) they grew up with, in the woods.

In the Oregon forests, they ran up against forest authorities, but have now reached an uneasy peace with these authorities. Legislation, Anna explains, requires "public forests" to be thinned for fire protection for one mile around all private structures, mandates the hiring of private companies to do the thinning, and allows logging but not mushroom picking in these "public forests."

The authorities have to work around these unwieldy and absurd laws. (You can be sure the anti-environment, pro-privatization Trump will make this situation unimaginably worse). Impromptu supply chains then emerged to connect the pickers with Japanese companies looking to sell to their domestic clients. Buyers, sorters, aggregators, and other intermediaries evolved, again mostly along traditional ethnic lines, according to the specialized skills needed.

Japanese import companies, many of them based in Vancouver BC, were ready to alienate the new product and accumulate the salvage. Once in Japan, the mushrooms are re-sorted, because one of their primary uses in Japan represents an exit again from the industrial economy. Matsutake mushrooms are prized as gifts given in an economy built on relationships. They cement deals, honor rituals, and provide tokens of appreciation.

Gifts of this sort were once hand-made, rather than dependent on the industrial economy. With the collapse of the industrial economy, which now concentrates wealth and power without adding any value (affixing a label and marketing do not add real value), we will have to learn to hand-make or hand-grown, our gifts again. The industrial economy, Anna explains, is unable to deal with limits. It requires an endless supply of controllable, manageable 'resources' and infinite growth.

As we reach the borders to increase, the corporations in the oligopolistic industrial economy have scrambled to outsource what they cannot control, and they use their control of supply chains to bottleneck the salvage and the gift economy, so they are forced to deal with, and through, the oligopolies (despite their massive cost).

One of the critical concepts of the book is the idea of precarity — the reality that 'natural' resources and events are unpredictable and uncontrollable and can be disruptive. "Precarity once seemed the fate of the less fortunate," she writes. "Now, it seems that all of our lives are precarious, even when, for the moment, our pockets are lined." The hallmark of a culture and world in collapse is that precarity is ubiquitous. Thanks to precarity, America cannot be "made great again," if it ever was, despite many Americans' nostalgia for that dream.

The economy of the future will be one of increasing precarity, leaving less and less room for the industrial economy as it grows increasingly unsustainable. We have to start imagining how this emerging post-industrial economy will emerge and how we can survive and thrive in it. That post-industrial economy will not be a knowledge economy, it will be principally a salvage economy, with elements of a gift and relationship economy. It can't be designed or controlled.

It will evolve as it must, as it always has, in patchy, unorganized, and then self- organized ways. It will be one, in Anna's words, of "disturbance-based ecologies in which many species [and cultures] sometimes live together without either harmony or conquest." It will take more imagination than what we have shown so far (Mad Max, neo-survivalism, and new old west scenarios) to navigate our way to such an economy collectively.

Such an economy cannot be imposed or centralized because it doesn't scale. Anna explains that the industrial economy began with European plantations, where (slave) labor and (conquered) land was, for a time, unlimited and controllable.

The model was then exported to the factory. Without infinite, controllable 'inputs,' the model comes undone, as has now happened.

The emerging salvage/gift/relationship economy will of necessity be local and opportunistic, responsive to what is available at hand. The yellow area of the chart above will gradually disappear, as we find we can no longer afford to allow capitalists to exploit their concentrated power to appropriate obscenely disproportionate wealth while doing nothing of value.

As they have disintermediated, so they will be disintermediated. Imports andexports will quickly become prohibitively expensive and rare, so production and consumption will mainly occur locally, near each other. That will require much more knowledge of local ecology and sustainable 'disturbances' that indigenous cultures had mastered. But that knowledge and those practices are local, and will not easily translate to areas they're most needed areas where indigenous knowledge has long been lost.

The idea of inventorying to tie up supply and force up prices will vanish — locally. It will not be tolerated. We will have to relearn how to accumulate only the salvage we need and how to do so and sustainably. And we will rediscover the value of gifts and relationships. The concept of 'property' will eventually die.

In the latter part of the book, Anna talks about the increasing importance of 'noticing' studying humbly how things appear to work and how small disturbances work or don't. She also explains how interdependent we are with other creatures. For example, pine wilt nematodes, which co-evolved with pines in North America and which take out only sickly trees (healthy trees are immune), traveled with American pines shipped to Japan in the last century.

Japanese pines have no immunity and hence were devastated by these insects, contributing to the scarcity of matsutake mushrooms in Japan and the explosion of matsutake picking in North America.

Nature selects relationships, rather than species, she asserts. Survival of the fittest entails how one fits in with one's fellow species in each local place, and that's more about relationships with other creatures (what one offers to the whole) than a competitive advantage. These, in turn, are the result of what she calls "contingencies of encounter," and these occur regularly and need to be observed and appreciated to thrive in any local environment.

She also speculates that, based on some recent research, viruses may be how DNA try out variations to see if they are naturally selected; they may not merely be 'random.' There is no such thing as a separate culture, she argues creatures of different species and their environments "co-culture" each other as much or more than beasts of a single species do. Cats and dogs and even aliens in industrial confined animal operations are our co-cultures.

The creatures in such services create a human culture of enduring willful ignorance, denial, and indifference to suffering that permeates far more than just what we choose to eat. So, rather than studying cultures in isolation, we should be studying assemblages, the contingent, precarious, endlessly disturbed, coalescing "polyphonic performances of living."

But we seem a long way from doing that, she concludes. Near the end of the book, she describes the modern Japanese phenomenon of hikikomori: "a young person, usually a teenaged boy, who shuts himself in his room and refuses face-to-face contact. Hikikomori lives through electronic media. They isolate themselves through engagement in a world of images that leave them free from embodied sociality — and mired in a self-made prison. They capture the nightmare of urban anomie for many; there is a little bit of hikikomori in all of us."

Magic Mushrooms

Mycology, the study of mushrooms, is bringing new admirers to the 'fungus among us." Already being used for a variety of medical reasons around the world, the little toadstool may be thrust into the spotlight soon as a successful alternative treatment for some stubborn imbalances.

Vegetarians value mushrooms due to their high nutritional value. They can produce vitamin D when exposed to sunlight. Mushrooms contain B vitamins, vitamin C, potassium, phosphorus, calcium, sodium, and zinc.

Medicinal mushrooms have thousands of compounds and nutrients that are health-strengthening. Eastern medicine, especially traditional Chinese practices, has used mushrooms for centuries. In the U.S., studies were conducted in the early '60s for possible ways to modulate the immune system and to inhibit cancerous tumor growth with extracts.

Mushroom hunting is popular, but it is not safe. Some edible mushrooms are almost identical to poison ones. It takes an expert to tell the difference. Also, fungi behave like a sponge and easily absorb toxins from soil and air. However, mushrooms are easily considered 'health food.'

Without the process of photosynthesis, some mushrooms obtain nutrients by breaking down organic matter or by feeding on higher plants. Another sector attacks living plants to consume them. Edible and poisonous varieties are found near roots of oak, pine, and fir trees.

Mushrooms were used ritually by the natives of Mesoamerica for thousands of years. They were widely consumed in religious ceremonies by cultures throughout the Americas. Cave paintings in Spain depict ritualized ingestion dating back as far as 9000 years. Psilocybin use was suppressed until Western psychiatry rediscovered it after World War II.

The controversial area of research is the use of psilocybin, a naturally occurring chemical in certain mushrooms. Psilocybin is effective in treating addiction to alcohol and cigarettes. New studies show that hallucinogenic drug might relieve anxiety and depression in some cancer patients. Mood raising effects that lasted at least several weeks after consuming the fungus were reported in some studies.

While fungus has fascinated people for centuries, it may finally be coming into a new era where its healing powers and unknown qualities are being discovered. The mushroom might very well hold the key to some long-ago locked mysteries and diseases. Medicinal use of mushrooms has been going on for thousands of years with good reason: they are useful. It is time for more focused research exploring new applications and powers of this delicate gift from nature.

➢ Drug Addiction

Magic Mushrooms, as they are known, are naturally occurring Fungi which are usually consumed raw or dried and ground up and drank in tea or coffee, and produce hallucinogenic effects. There are many, many different types and varieties of magic mushrooms with varying strengths.

The mushrooms free up the imagination to internal or external influences and let it run without bounds, whether the 'trip' be pleasurable or a nightmarish experience is almost uncontrollable. It generally takes no longer than an hour for the trip to engage, and can last up to 6 hours. It is like a less intense alternative to the far more dangerous semi-synthetic hallucinogen LSD.

While the long term effects of taking magic mushrooms regularLy are somewhat unknown, the biggest problem is their natural availability (they grow in wild grazing fields in or around cow and horse feces). This can be somewhat of an irresistible lure to the thrill-seeking mushroom users who'll go out and collect them on their own, thinking every mushroom is consumable.

However, not all of these fungi are the desired ones, and it can be tough though to distinguish ones who are or aren't toxic. Some of these mushrooms are highly poisonous and can kill in a prolonged and painful way, for example, fever, vomiting, and diarrhea. Some even have a delayed reaction taking days to show any signs or symptoms before taking your life with absolutely no antidote.

Because Magic Mushrooms are naturally occurring and not 'processed' in any way before consumption, they are somewhat naively considered a safe drug.

Absolutely no prescription is secure, and most medications are naturally occurring or refined from natural plants or fungi anyway. They aren't known as an addictive or heavy drug, nor are they as violent or psychologically damaging as LSD, nor are they socially corroding such as crack or heroin.

Depending on the mushroom-users mental predisposition; however, mushrooms can have a damaging effect on the user. For instance, if the user is prone to having a fragile mental state or is very suggestible, they may believe their hallucinations to be the manifestation of something real and become somewhat obsessed with it and damaged by it.

One such documented case of these extremities involved a young man who began taking mushrooms and started having the recurring hallucination of a flower dressed up as a court-jester, which repeatedly taunted him with scarring insults. As unusual as it sounds, without discounting these experiences merely as hallucinations, he believed this abusive-flower to be the manifestation of truths about himself and spiraled into a severe depression. He and his friends admitted he was beautiful before taking mushrooms, but somewhere during the course, a can of worms was opened for him.

Sadly, to this day, he still struggles with emotional and mental issues, which weren't there before the advent of his life-changing hallucinations. It would be impossible to say for sure in such a case if the mushrooms were responsible for triggering such continuing mental problems, or an underlying mental illness was already present, and the mushroom use was inconsequential, but it is always worth bearing in mind.

The Type Of Magic Mushroom You Should Consume

The type of magic mushroom you take, and how much of it you consume will have a profound effect on your trip. For example, the red and white-capped Amanita muscaria is a magic mushroom that offers its own unique experience, different from any other magic mushroom. The ergot mushrooms (more scientifically called "Claviceps purpurea") provide a psychedelic experience similar to LSD (which has been synthesized initially from this mushroom). From a shamanic perspective, the different species would be considered to be different "teachers," each with their personality and wisdom to impart.

The Most Commonly Consumed Psychedelic Mushroom Species Are:

- **Panaeolus:** Subbalteatus, Tropical
- **Psilocybe:** Sarcocystis, Caulescent, Cubensis, Cyanescens, Mexicana, Pellicular, Semilanceata, Stunts
- **Copelandia:** Cyanescens, Cambodgeniensis

These Species, Popular Varieties Include:
- Psilocybe cubensis
- Sclerotia also known as philosopher's stones, a truffle species of psychedelic mushrooms
- Copelandia cyanescens

Everyday Experiences Of Magic Mushroom Trips

Now that you understand what factors will influence your magic mushroom trip, what is the experience actually like? Every person's magic mushroom trip is unique to that individual, due to the myriad internal factors that affect one's knowledge and interpretation of reality, regardless of the species of mushroom, you consume. However, in general terms, you can expect to experience the following phenomenon:

> **Visual Enhancement**

The visuals may vary. Some individuals may experience more external visions, while others might perceive more intimate images. For some, ideas may be light, but the journey filled with cosmic downloads. Typically, mushrooms will make your experience of the world more vivid. Colors are brighter, and if you are walking on grass, the Earth will appear to be abundantly alive...in fact, the ground might look like it's breathing.

You might see trails as birds fly across the sky, and as people, as well as other objects, move. You might see faces in trees and rocks. If you've eaten a lot of magic mushrooms, and if they are the highly visionary species, you may see things that will make you rub your eyes and look again. We were walking over a meadow at an outdoor concert in the mountains. We had eaten way too many mushrooms.

➤ Auditory Hallucinations

Sounds may appear enhanced and somewhat distorted, in highly comical ways. Some kinds of music sound amazing on magic mushrooms. Others, not so much. Experiencing music on psychedelics may permanently alter your musical tastes.

➤ Distortions In Time & Space

Many mushroom trippers will experience time slowing down or standing still. This may be because mushrooms have a way of making you extremely present, and the more present you are, the more aware of the time you can be.

Distances may also appear to be warped, objects looking way closer or further away than usual. You may also have a vast experience of space-time, calling into question our society's belief in time being linear, and space being "out there."

➢ Emotional Release

Psilocybin mushrooms can heal trauma by allowing the user to revisit a traumatic experience from a different perspective. What comes with this experience is often a cathartic emotional release Through hysterical laughing, crying, and so on – giving way to the lightness of being and a new sense of self. Current scientific studies include the use of psilocybin mushrooms in the treatment of depression, PTSD, and end of life anxiety.

➢ Ontological Paradigm Shifts

Magic mushrooms have a remarkable way of making bullshit belief systems visible. I suspect that's why authorities are terrified of people using mushrooms – not because they are dangerous to the individual but more because they are hazardous to those in power. If magic mushrooms were legal, you'd have tens of thousands more people rising to the control systems in place to keep humanity meek and compliant.

A collective magic mushroom experience is questioning the validity of your long- held beliefs, and an openness to try on new ideas. Part of the medicine of the mushroom is its ability to show you what's really, truly important. It's not a nice car, or a good job, or lots of money. It's usually so much simpler...Life. Health.

Family. Love.

➤ Body Bliss

Magic mushroom users often report the physical sensation of being euphorically blissed out. Bodily sensations are pleasurably heightened, doing yoga, dance, and enjoyable massage activities to experience on magic mushrooms. For the first time, ecstasy took on real meaning. ~ R. Gordon Wasson

➤ Cognitive And Spiritual Rebirth

After going through a complete re-examination of your belief system, personal history, and experiencing the miracle of being alive freshly and differently, you may feel a new kind of appreciation for everything you see, smell, or touch.

It is like you are a new-born baby again, exploring the world through your untethered senses and with an expanded mind. You may come to understand your behavior, your role in the world, and your surroundings in a new, more integrative way.

A study done at Johns Hopkins University revealed the extraordinary spiritual power of magic mushrooms. In interviewing the participants after the study, a third of them said their experience was among the top five most meaningful of their lives.

A third-ranked it in the top five most spiritually significant experiences of their lives while 62% of them stated that the experience was among the top 10 most difficult ones in their lifetime.

➤ The Unexpected

Magic mushrooms are magical for a reason. Eating them tends to bring on the unexpected synchronicities and experiences that seem to have a profound, life-changing implication for you. More importantly, they tend to connect us with something precious that many of us forget, lose touch with, especially as we grow older, and life's responsibilities weigh us down. And that is the magical wonder of life.

Healing Properties Of Mushrooms

If you believe the ancient Chinese medical sciences, specifically those dating back to 2000 years, they specify certain species of mushroom that possess healing properties. These species are also believed to promote physical and mental health and longevity. In ancient times, these rare herbs were available only to the ruling clan of classical China, but not anymore!

There are different varieties of mushrooms such as Reishi, Cordyceps, or Shitake, which were available to the highest clan of the society, and were treated as "The Emperor's Secret." Royal families could only afford them due to their scarcity and cost. They believed that mushrooms promote health, vigor, longevity, and immortality.

Scientists of today's generation seem to agree with those of ancient times. In recent findings, it is confirmed that mushroom nutrition prevents you from many health diseases such as tumors, cancer, fatigue, viral infection, and Hepatitis B.

Other than this, there exist numerous traditional varieties of mushroom that are significantly helpful in boosting immunity, and prevent you from various diseases such as common cold, influenza, inflammation, arthritis, allergy, chronic Bronchitis, and herpes.

If you want to improve your overall health and wellbeing, or boost your ability or fight various types of chronic disease, including cancer, you must include mushrooms in your daily diet to get essential health benefits. You can use mushrooms in soups, sandwiches, and salads to enjoy their health benefits.

Mushrooms contain various minerals such as Copper, which has proven the ability to fight multiple health disorders. The Potassium present in mushrooms helps in preventing various cardiac problems, including heat strokes.

Mushrooms contain more potassium than bananas. Apart from adding flavor, taste, and variety in your food, mushrooms provide you an essential cover against several diseases. Therefore, it is a good idea to include mushrooms in your diet.

New Chapter Mushroom Nutrition And Beta-Glucan Polysaccharides

Long known for their healing and health-related properties, mushrooms are grown organically to promote their qualities and keep them pure. Glucans are polysaccharides that contain only glucose as a part of their structural components. New Chapter mushrooms are unlike any other mushrooms anywhere. Organically grown for medicinal purposes, they are meant to increase aspects of the immune system.

They contain antioxidants like ascorbic acid, phenolic compounds, tocopherols, and carotenoids. Various strains of mushrooms are grown for their antioxidant activity, among which include the maitake, reishi, oyster mushrooms, and shiitake.

➤ Medicinal Mushrooms And Blood Sugar Levels

Through much research, it has been shown that some of the medicinal mushrooms, such as the reishi, Agaricus campestris, and Chaga, are capable of lowering elevated blood sugar levels. The maitake is especially suited for lowering blood sugar because it contains a compound that is known to be an alpha-glucosidase inhibitor.

➤ Medicinal Mushrooms And Cholesterol

The reishi and Agaricus blazei mushrooms have shown us to be capable of effectively inhibiting cholesterol levels. Shiitake mushrooms contain a specific the anti-cholesterol compound, which is known as eritadenine and oyster mushrooms, are found to contain lovastatin, which is well known to lower cholesterol naturally.

➢ Mushroom Research At New Chapter

Mushrooms go through various life cycles, each of which adds critical and unique nutrients. These, in turn, offer more excellent protection from harmful diseases and promote their healing qualities. Some proven facts about mushrooms include:

- After over fifty years of research, mushrooms have been confirmed to possess inherent therapeutic capabilities that address a wide range of health-related concerns.
- Mushrooms provide needed respiratory support while also promoting brain function and a healthy liver.
- Mushrooms have proven abilities to support the immune system.
- Medicinal mushrooms give Vitality, wellness, sexual function, and endurance a boost.
- Whole mushrooms, containing their entire life cycle, are essential for their health-promoting qualities.

New Chapter Vitamins and nutritional mushrooms are carefully researched and held up to the highest of standards to provide the best in health care while preventing diseases. Scientific as well as medical research, along with the traditional use over time by many cultures, provide proof that medicinal mushrooms have much to offer in the way of overall Vitality, liver and respiratory support, and immunity to diseases.

Discover As with anything, something potentially good can have devastating effects as well. Take, for example, mushrooms. Mushrooms have documented history attached to their beneficial uses, medicinally. Reishi Mushroom

Supplements contain one of the most widely used and well-known mushrooms for their medicinal purpose.

Mushrooms are part of the fungi family. Different kinds may have different effects on the body and mind. On one extreme, there are the skull-cap mushrooms, which can cause death. Then there are the kind people to have mind-altering high's. Then there are others like the Reishi, which are used for healing benefits.

The Reishi mushrooms, among one of the most popular Chinese mushroom, due to a long list of essential health benefits. This includes such properties, such as antioxidants. This property accounts for the reason the dried powder of the mushroom was used in ancient China as a chemotherapy agent.

Another name for Reishi is Ganoderma or Ling Zhi. A lot of research continues on mushrooms, such as the Reishi that are found as having many health benefits.

Some potential benefits include the prevention of some kinds of cancers. One study found that there may be properties of the mushroom, which prevent breast and prostate cancer cells from invading. Most studies found some cancer-fighting property. Anti-cancer, Anti-tumor, reduction of blood pressure, cardiovascular, Lyme disease, anti-viral, stress reducer, nerve tonic, lack of energy, diabetes, and Lung/respiratory strengthener are just a few of the many other potential healing properties.

When tested for potency in effectiveness when used, most found the actual mushroom worked best. Therefore, this suggests that some of the Reishi Mushroom Supplements may not work as well if not prepared correctly. It's always good to read the label and make sure to check with you're a doctor before starting anything new.

How To Find Mushrooms

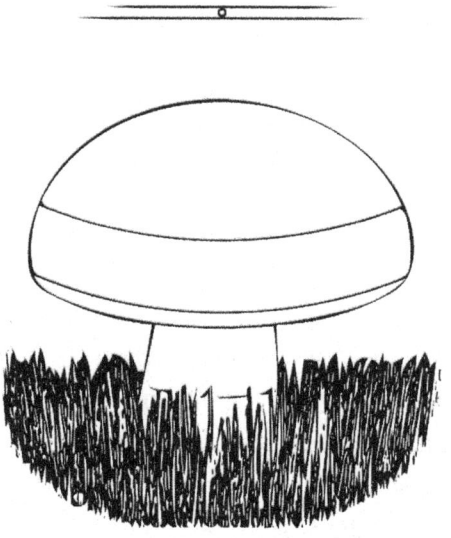

Your first bold move toward identifying mushrooms is to find them. Equip yourself with a pocket knife, a sack or backpack, and a roll of wax paper for wrapping specimens. Walk, ride your bike (our preferred method), or--if you must--drive your car around the neighborhood and see what you find.

In urban areas, mushrooms can grow anywhere, from your basement to the empty lot next door. Look in the grass, under bushes, along streams, on wood, in flower pots. There's no secret to it. Over time, however, you will learn which habitats your favorite mushrooms like and when they might grow, and you will look there first.

Here's the gist of how to do it: We know that oyster mushrooms want to grow in the spring (if it's been wet) on cottonwood stumps in my back yard. That's where we look first. Next, we check any stump we can find, particularly stumps from the kinds of trees that we know oysters like to grow on.

Most city mushrooms grow in lawns or flower beds with or without manure or wood chip mulch and on live trees or wood stumps. But other sites are also productive -- empty lots, under trees, and indoors in flower pots. Unkempt lawns, infrequently mowed, with weeds (spared from herbicides) are the most productive. Some common urban species tend to grow in or near disturbed areas.

For example, look for the urban mushroom (Agaricus torques) in hard-packed soil; the shaggy mane (Coprinus comatus) along the roadside; and puffballs near curbs. Other preferences appear more specific -- stinkhorns (Phallus impudicus) under lilac; shaggy parasol under spruce trees; and Japanese parasol (Parasol plicatilis) under hawthorn trees. Keep an eye out for the domicile cup fungus (Peziza domiciliary) growing in the basement on moist walls and carpets.

Combine your knowledge of where mushrooms like to grow with your insight on when they want to grow there. For example, most mushrooms grow in Denver in

April through September. The winter mushroom (Flammulina velutipes) appears earlier and survives later. The mica cap (Coprinus micaceous) is also most common in spring and fall. You're more likely to find the shaggy parasol (Lepiota rachodes) and the inky cap (Coprinus atramentaria) in the fall. The fairy ring mushroom is around all summer long.

Moisture, of course, is also crucial. Irrigation by the city and property owners stabilizes ground moisture, supporting mushroom growth even in dry periods. However, soaking rain will trigger the most intense fruitings. Look in the north- and east-facing lawns and gardens after a rain.

These areas, in the shadows of houses in the hottest times of the day, stay wetter more extended than the west and south-facing sides. Puffballs (e.g., Lycoperdon species) and their relatives are most resistant to drought. The fairy ring mushroom will dry out and then reconstitute following a rain.

➢ How To Pick Mushrooms

Once you've found mushrooms, you need to determine whether they are growing on private or public property. If it's closed, ask permission from the property owner to pick them. Most of the time, you'll be encouraged to remove as many as possible. (See discussion in Introduction.)

Pick mushrooms carefully to preserve the characteristics that you will need to identify the fungus later. Dig the mushroom out completely, making sure you extract the entire stem from the ground. Famous mushroom characters are sometimes found at the base of the stem.

COLLECT SAMPLES OF ALL AGES. DIFFERENT CHARACTERS EMERGE AS THE MUSHROOM GROWS.

Note where you found the mushroom and its habitat-- or write down this information on a label. Was it under trees? If so, what kind? Was it growing from grass? Was it growing in a partial circle or fairy ring? Roll up your specimen--with your label-- in a piece of wax paper and twist the ends as shown. This will protect it so you can study it in more detail later.

➢ How To Identify Mushrooms

When you're finished with your mushroom hunt, gather together and unwrap the mushrooms that you've found. It's best to have an experienced collector on hand to help you identify them. But a careful beginner with a couple of mushroom field guides can begin to identify mushrooms.

Examine your collections one at a time. There is no single rule that allows you to determine if a mushroom is edible. Similarly, it would be wrong to say that all white mushrooms are edible. Or all brown ones or all red ones. The only way to identify a wild mushroom is to know the characteristics of the mushroom that you are identifying.

The best way to avoid making mistakes is to know not only the mushroom you want but also mushrooms that look like your desired mushrooms. If you know your "lookalikes," you are less likely to misidentify mushrooms as a result of your failure to note a couple of critical characteristics.

➢ Smell It

Flowers aren't the only things that smell in the garden. Mushrooms have unusual smells, which can help with identification. The corn silk smell of some Inocybespecies takes some people back to their childhoods of eating fresh corn every day, all summer long. Don't miss one of the very distinct characteristics of a mushroom and find out where it can transport you.

Here Are A Few Mushrooms That You Can Sniff From Lawns And Gardens

- Agaricus xanthodermus: creosote or medicinal
- Agaricus Augustus: anise or marzipan
- Inocybe sp: corn silk or spermatic
- Phallus impudicus: fetid, to put it mildly
- Touch it

Not all mushrooms are the same to touch. They are fuzzy, slimy, dry, smooth, spiny, hairy, scaly, waxy, and more. It's important to note how the mushroom feels.

➢ Taste it

Once you've learned a bit about mushrooms, you can begin tasting them to help you identify them. For example, Russula emetica is intensely bitter. Take a small piece on the tip of your tongue, hold it there for a few seconds, and then spit it out. If you spit it out, it won't hurt you.

➢ Make A Spore Print

Mushroom spores come in all colors from white and black to pink and purple. Determining the color of a mushroom's spores can help you identify the fungus. Even though spores are microscopic, you can frequently figure out their color by making a "spore print." Most city mushrooms produce spores on gills, which are the blade-like structures on the underside of a mushroom's cap. To make a spore print, place the mature mushroom, gills facing down, on a white piece of paper.

Cover it and leave it for a couple of hours, and you may find a beautiful--and delicate--spore print.

Mushrooms that don't have gills produce spores in other structures. A puffball, which starts as a solid white mass, slowly dries out, finally puffing out dust-like spores when it is squeezed or disturbed. Some mushrooms produce spores in "pores," which appear under the cap instead of gills. Other mushrooms make spores on "teeth," spine-like structures under the cap.

➤ Look For A Cup, A Ring Or Warts

In addition to producing spores of different colors, mushrooms grow other structures that provide clues to their identities. For example, the diagram below illustrates the development of mushrooms in the genus Amanita. It starts (in the figure below) as an "egg" or "button," covered with a "universal veil." When it

emerges from the button (in number two), the remains of the universal veil leave the "cup" or "volva" at the base and the "patches" or "warts" on the top of the cap.

At this stage, a "partial veil" connects the cap to the stem, covering the gills. When the cap expands from the stem (in figure 3), the gills become visible, and the remains of the partial veil form a "ring" on the stem. If a mushroom that you've found has these structures-- and other information is consistent--you may conclude that you've seen an Amanita. But you must be thorough because other gilled and non-gilled mushrooms may have these structures, such as a ring or cup, as well.

➢ Look At The Shape Of The Cap

Mushroom caps come in many different shapes. You should look carefully at the Mushroom's cap during various stages of development. Some young mushroom caps may be conic of convex, later becoming plane or depressed. Also, be on the lookout for slight variations in cap shape, such as a "knob" or "umbo" on top.

➢ Look At How The Gills Attach To The Cap

A mushroom's gills--if it has gills--attach to the stem in several ways. Some gills are not attached to the stem at all. These are called "free" gills. Others are decurrent, running down the stem. It can sometimes be challenging to determine precisely how the gills attach to the stem without looking at specimens of varying ages.

➢ Look At The Shape Of The Stem

Mushroom stems are shaped in distinctive ways, from "bulbous to "equal." Again, it's best to look at several specimens

> **Look At How The Stem Emerges From The Cap**

The stem of a mushroom attaches to the cap in a variety of ways. In some cases, of course, there is no stem at all--or no cap for that matter. In other cases, the stem comes from the center of the cap.

> **Look At The Colors And Markings**

When you start to look at individual mushrooms in more detail, the amount of fantastic stuff to see expands exponentially. And all of it can provide clues to the Mushroom's identity. For example, mushrooms can vary in color from all shades of brown to all shades of red. They can disintegrate into an inky mess. They can have striated or smooth caps.

They can have scaly, dotted, or hairy stems. Even the ring on the stem can have distinctive--and beautiful--forms. The best mushroom identifiers hone their skills of observation, allowing them not only to classify mushrooms more accurately but to admire these unbelievable organisms more deeply.

> **Look Through A Microscope**

This website aims to help you identify mushrooms in the field under the swing set if necessary. Many mushrooms can be identified adequately wherever you find them with the help of this field guide (and perhaps a couple other for cross- checking), mainly if specimens of varying maturities are available.

However, it is impossible to identify many mushrooms with certainty without checking microscopic characteristics. So, if you get serious about trying to put a complete name on all mushrooms you find, you'll have to learn how to handle a microscope. If you do, you'll find a new world of spore shapes (spiked, ribbed, lumpy, spherical, elliptical), reactions, and colors to observe. You'll also have to buy another mushroom field guide because this one does not cover microscopic characters

A Simple Guide To Common Mushrooms

Portobello, Shitake, White Cap, Oyster. These are just a few of the many species of mushrooms available in your market. Right that you buy most mushrooms in the market, but would it give you greater pleasure when you grow them yourself?

Looking at those white globs would make you think that it might be too hard to grow, much more cultivate mushrooms. But then again, looks can be deceiving. And it sure does! Mushrooms are one of those things that can surely capture your imagination. They may seem delicate to look at, but no expert hand is needed to grow these babies.

What Kind Of Mushroom Do You Want To Grow?

This is an essential part of the process if you want to cultivate your mushroom garden. The reason why you need to know the type of Mushroom is each Mushroom has different needs. Some mushrooms are better used as dowels, and others need to spawn. But for beginners, it is suggested that you use mushroom growing kits so it would be easy to do, plus, this can become a small project over autumn or spring.

Another factor that you need to consider is the substrate where you are going to "plant" your mushroom spawns.

➢ Log Substrate

Some mushrooms are better planted on logs such as willow, oak, and beech.

These include the oyster mushroom and the shitake mushroom. You do not plant directly onto the woods. You need dowels for that. These dowels are saturated with the mycelium, the part of the Mushroom from were the mushrooms that we eat spring out.

These dowels are then inserted into the log, sealed with some wax to prevent any contamination, stored in a dark and dry place until the mushrooms are ready to be harvested. The records should be protected from direct sunlight as well as strong winds. So, better wrap them in a black polythene bag. It would take around a year and a half for the fungi to colonize the entire log.

➢ Straw Substrate

Perhaps one of the most comfortable substrates to deal with is the straw. First, you need to sterilize the straw. Do this by pouring boiling water on the straw.

Once the straw has cooled, sprinkle the spawn, which is included in the mushroom kit. Make sure that you shake the bag with the straw and the spawn to spread the spawn evenly. Seal the bag. Place this is a dark and moist place. The mushrooms can be harvested in about six weeks.

➢ Growing Mushrooms Outdoors

This is one of the traditional ways of growing mushrooms. You use grain spawn to inoculate the mushrooms into the manure or soil substrate.

Delve Deeper Into The History Of Mushroom

Mushroom is historically believed to have existed with its enhancing response. This is mainly due to its higher content of complex sugars, Polysaccharides, and a whole lot more. This has its double-direction effect brought to the immune system. This thereby helps in the regulation of antibody production in the entire body system. Mushroom, as such, is recognized for its medicinal marvel and spiritual potency.

By Increasing The Immune System, Certain Mushrooms Can Indirectly Help In Many Areas, Including:

- Beneficial in Cancer Treatment
- Inhibits Bacteria
- Reduces the fat levels of the blood
- Lowers hypertension and high blood pressure
- Provides liver regeneration and protection and improves liver function
- Provides its Anti-Arthritic Relief
- Alleviates Allergies and Inhibits Histamine Release
- Helps Contribute to the Relief of Pain

- Relieves Tensiontension due to Stress and Relieves Insomnia
- And much more
- medical stethoscope

➢ **Mushroom Of Immortality**

With all these benefits of mushrooms, it is just right to believe that Mushroom, like Reishi, is of immortality. Ancient people have known this ever since and have found it to be useful in Conventional Chinese Herbalism practices. This also positively brings its colossal impact to the heart Qi, strengthens the intelligence, and cures forgetfulness.

One impressive thing to take note of mushrooms is that it relieves the causes of diseases and illnesses, especially stress. Even the Taoists claimed that the Reishi mushroom promotes centeredness, calmness, inner strength, and emotional awareness. They also have utilized to enhance meditative practices and to protect the spirit, mind, and body. This way, you could obtain a healthy and long life based on spiritual immortality.

Reishi exactly promises longevity to human beings. Especially with Traditional

Chinese Herbalism, they considered mushroom medicine as it helped ease

Tensiontension, calm the mind, strengthen the nerves, sharpen the concentration levels and improve a sense of focus, build wisdom and build will power.

Now, you have learned more about medicinal Mushroom, including the history of mushrooms used for healing!

➢ Mushrooms Healing Properties – Things To Consider

Now, the truth is that if you want to live a healthy life and have a robust immune system, then you would have to take quite a few things into proper account.

Nutrition, of course, is without a doubt amongst the most important ones. With this in mind, we would like to inform you on the incredibly strong healing Properties of mushrooms. As it turns out for thousands of years, these incredible fungi can do you quite a lot of good. So, let's go ahead and take a look.

➢ The Powers Of The Reishi Mushroom

This is a type of mushroom that has been used by different physicians for centuries, especially relevant in China. Consequently, ancient Chinese physicians determined that the mushroom could prevent a lot of illnesses, and it can cure a lot of different diseases. This is the main reason for which it got the name

"Mushroom of Immortality." That's right – the Reishi is known for increasing longevity and help you live a healthy life.

As it turns out, the primary source of its medicinal and healing properties is due to a carbohydrate complex. A scientific name for this complex is polysaccharides.

➢ The Turkey Tail

Turkey Tail is another particularly beneficial mushroom that is going to contribute to a healthy lifestyle. It is abundant throughout the majority of the forests in North America, and it usually resides amongst felled trees. Now, the mushroom is used to create particularly prominent extracts that are capable of helping with the fight against cancerous conditions. It is also capable of repelling viruses that are going to set the root for cancer.

There are quite a lot of different researches that attest to those above medicinal and healing properties of the mushrooms that we mentioned. For instance, there was a research conducted in the University of Minnesota as well as in the Bastyr University, both of which determined that the Turkey Tail mushroom strengthens the immune system of the patients, hence helping them fight off different cancerous conditions and helping out conventional cancer therapies.

Of course, there are quite a few other types of mushrooms apart from the Reishi and the Turkey Tail, which can promote longevity. Lion's Mane, Maitake, Chaga, Psilocybin mushrooms as well as many more of the kind is capable of helping you out in the quest to sustain a healthy lifestyle. This is important, and it's going to ensure that you are energetic and enthusiastic.

The Grower's Guide To Psilocybin Mushroom

More and more people are growing psilocybin mushrooms at home. As well as providing a reliable, year-round supply, home cultivation eliminates the risk of misidentifying mushrooms in the wild. For many growers, it's also a fun, low-cost hobby.

If you don't know how to grow mushrooms at home, you may be tempted to start with a psilocybin mushroom grow kit. These ready-to-use packs contain a living mycelium substrate (the material underlying mushroom growth) that, in theory, you need to keep humid.

In reality, you're better off starting from scratch. Making your substrate is not only more consistent but, if you do it right, it should be less prone to contamination as well. There's also not a massive difference in price, and you'll end up learning a lot more.

(If you're interested in legally consuming psilocybin, you may join Third Wave founder Paul Austin for legal psilocybin retreats in Amsterdam.)

This guide is based on Robert "Psilocybe Fanaticus" McPherson's eponymous PF Tek—the method that revolutionized growing mushrooms indoors. McPherson's key innovation was to add vermiculite to a grain-based substrate (as opposed to using grain alone), giving the mycelium more space to grow and mimicking natural conditions.

Although his method is a little more labor-intensive than others, often for a lower yield, its simplicity, low cost, and reliability make it ideally suited to beginners. It also makes use of readily available materials and ingredients, many of which you may already have.

➢ Spore Syringes

The one thing you might have trouble getting is a right spore syringe. This will contain your magic mushroom spores and be used to "sow" them into the substrate. Some growers have reported issues of contamination, misidentified strains, and even syringes containing nothing but water. However, as long as you do your research and find a reputable supplier, you shouldn't have any problems.

In any case, after you've grown your first batch (or flush) of mushrooms, you can start filling syringes of your own.

➤ What Variety Should You Choose?

As you learn how to grow mushrooms indoors, you'll want to decide on a species and strain. Most suppliers offer a range to choose from, but the Psilocybe cubensis B+ and Golden Teacher mushrooms are among the most popular for beginners. While not as potent as some others, like Penis Envy, they're reportedly more forgiving of sub-optimal and changeable conditions.

What you will need

➤ Ingredients

- Spore syringe, 10-12 cc
- Organic brown rice flour
- Vermiculite, medium/fine
- Drinking water

➤ Equipment

- 12 Shoulderless half-pint jars with lids (e.g., Ball or Kerr jelly or canning jars)
- Hammer and small nail
- Measuring cup
- Mixing bowl
- Strainer
- Heavy-duty tin foil
- Large cooking pot with a tight lid, for steaming
- Small towel (or approx. 10 paper towels)
- Micropore tape
- Clear plastic storage box, 50-115L
- Drill with a ¼-inch drill bit
- Perlite
- Mist spray bottle

➤ Hygiene Supplies

- Rubbing alcohol
- Butane/propane torch lighter
- Surface disinfectant
- Air sanitizer
- Sterilized latex gloves (optional)
- Surgical mask (optional)
- Still air or glove box (optional)

> **Instructions**

The primary PF Tek method is pretty straightforward: Prepare your substrate of brown rice flour, vermiculite, and water, and divide it between sterile glass jars.

Introduce spores and wait for the mycelium to develop. This is the network of filaments that will underpin your mushroom growth. After 4-5 weeks, transfer your colonized substrates or "cakes" to a fruiting chamber and wait for your mushrooms to grow.

NOTE: Always ensure good hygiene before starting: spray an air sanitizer, thoroughly disinfect your equipment and surfaces, take a shower, brush your teeth, wear clean clothes, etc. You don't need a lot of space, but your environment should be as sterile as possible. Opportunistic bacteria and molds can proliferate in conditions for cultivating shrooms, so it's crucial to minimize the risk.

STEP 1: PREPARATION

Prepare Jars: With the hammer and nail (which should be wiped with alcohol to disinfect) punch four holes down through each of the lids, evenly spaced around their circumferences.

Prepare Substrate: For each jar, thoroughly combine ⅔ cup vermiculite and ¼ cup water in the mixing bowl. Drain excess water using the disinfected strainer.

Add ¼ cup brown rice flour per half-pint jar to the bowl and combine it with the moist vermiculite.

Fill Jars: Being careful not to pack too tightly, fill the jars to within a half-inch of the rims. Sterilize this top half-inch with rubbing alcohol top off your jars with a layer of dry vermiculite to insulate the substrate from contaminants.

Steam Sterilize: Tightly screw on the lids and cover the pots with tin foil. Secure the edges of the foil around the sides of the jars to prevent water and condensation from getting through the holes.

Place the small towel (or paper towels) into the large cooking pot and arrange the jars on top, ensuring they don't touch the base.

Add tap water to a level halfway up the sides of the jars and bring to a slow boil, ensuring the jars remain upright.

Place the tight-fitting lid on the pot and leave to steam for 75-90 minutes. If the cup runs dry, replenish with hot tap water.

NOTE: Some growers prefer to use a pressure cooker set for 60 minutes at 15 PSI.

Allow To Cool: After steaming, leave the foil-covered jars in the pot for several hours or overnight. They need to be at room temperature before the next step.

STEP 2: INOCULATION

Sanitize And Prepare Syringe: Use a lighter to heat the length of your syringe's needle until it glows red hot. Allow it to cool and wipe it with alcohol, taking care not to touch it with your hands. Pull back the plunger a little and shake the syringe to distribute the magic mushroom spores evenly.

NOTE: If your spore syringe and needle require assembly before use, be extremely careful to avoid contamination in the process. Sterilized latex gloves and a surgical mask can help, but the surest way is to assemble the syringe inside a disinfected still air or glove box.

Inject Spores: Remove the foil from the first of your jars and insert the needle as far as it will go through one of the holes. With the needle touching the side of the jar, inject approximately ¼ cc of the spore solution (or slightly less if using a ten cc syringe across 12 jars).

Repeat for the other three holes, wiping the needle with alcohol between each. Cover the holes with micropore tape and set the jar aside, leaving the foil off.

Repeat the inoculation process for the remaining jars, sterilizing your needle with the lighter, and then alcohol between each.

STEP 3: COLONIZATION

Wait For The Mycelium: Place your inoculated jars somewhere clean and out of the way. Avoid direct sunlight and temperatures outside 70-80 °F (room temperature). White, fluffy-looking mycelium should start to appear between seven and 14 days, spreading outward from the inoculation sites.

NOTE: Watch out for any signs of contamination, including strange colors and smells, and dispose of any suspect jars immediately. Do this outside in a secure bag without unscrewing the lids. If you're unsure about whether a pot is contaminated, always err on the side of caution even if the substrate is otherwise healthily colonize as some contaminants are deadly for humans.

Consolidate: After three to four weeks, if all goes well, you should have at least six successfully colonized jars. Leave for another seven days to allow the mycelium to strengthen its hold on the substrate.

STEP 4: PREPARING THE GROW CHAMBER

Make A Fruiting Shotgun Chamber: Take your plastic storage container and drill

¼-inch holes roughly two inches apart all over the sides, base, and lid. To avoid cracking, drill your holes from the inside out into a block of wood.

Set the box over four stable objects, arranged at the corners to allow air to flow underneath. You may also want to cover the surface under the table to protect it from moisture leakage.

NOTE: The fruiting shotgun chamber is far from the best design, but it's quick and easy to build and does the job well for beginners. Later, you may want to try out alternatives.

Add Perlite: Place your perlite into a strainer and run it under the cold tap to soak.

Allow it to drain until there are no drips left, then spread it over the base of your grow chamber. Repeat for a layer of perlite roughly 4-5 inches deep.

STEP 5: FRUITING

"Birth" The Colonized Substrates (Or "Cakes"): Open your jars and remove the dry vermiculite layer from each, taking care not to damage your substrates or "cakes" in the process.

Upend each jar and tap down onto a disinfected surface to release the cakes intact.

Dunk The Cakes: Rinse the cubes one at a time under a cold tap to remove any loose vermiculite, again taking care not to damage them.

Fill your cooking pot, or another large container, with lukewarm water and place your cakes inside. Submerge them just beneath the surface with another cup or similar heavy item.

Leave the pot at room temperature for up to 24 hours for the cakes to rehydrate.

Cubes: Remove the cakes from the water and place them on a disinfected surface.

Fill your mixing bowl with dry vermiculite.

Roll your cakes one by one to thoroughly coat them in vermiculite. This will help to keep in the moisture.

Transfer To Grow Chamber: Cut a tin foil square for each of your cakes, large enough for them to sit on without touching the perlite.

Space these evenly inside the grow chamber.

Place your cakes on top and gently mist the chamber with the spray bottle. Fan with the lid before closing.

Optimize And Monitor Conditions: Mist the chamber around four times a day to keep the humidity up, taking care not to soak your cakes with water.

Fan with the lid up to six times a day, especially after misting, to increase airflow.

NOTE: Some growers use fluorescent lighting set on a 12-hour cycle, but indirect or ambient lighting during the day is beautiful. Mycelium only needs a little light to determine where the open air is and where to put forth mushrooms.

STEP 6: HARVESTING

Watch For Fruits: Your mushrooms, or nuts, will appear as tiny white bumps before sprouting into "pins." After 5-12 days, they'll be ready to harvest.

Pick Your Fruits: When available, cut your mushrooms close to the cake to remove. Don't wait for them to reach the end of their growth, as they'll begin to lose potency as they mature.

NOTE: The best time to harvest mushrooms is right before the veil breaks. At this stage, they'll have light, conical-shaped caps and covered gills.

Storage

Psilocybin mushrooms tend to go bad within a few weeks in the fridge. So if you plan to use them for microdosing or you want to save them for later, you'll need to think about storage. The most effective method for long-term storage is drying.

This should keep them potent for two to three years as long as they're kept in a cool, dark, dry place. If they're stored in the freezer, they'll pretty much last indefinitely.

The lo-fi way to dry your mushrooms is to leave them out on a sheet of paper for a few days, perhaps in front of a fan. The problem with this method is they won't get "cracker dry." That is, they won't snap when you try to bend them, which means they'll still retain some moisture.

They may also significantly diminish in potency, depending on how long you leave them out. Using a dehydrator is by far the most efficient method, but those can be expensive. A good alternative is to use a desiccant as follows:

Air dry your mushrooms for 48 hours, ideally with a fan.

Place a layer of desiccant into the base of an airtight container. Readily available desiccants include silica gel kitty litter and anhydrous calcium chloride, which you can purchase from hardware stores.

Place a wire rack or similar set-up over the desiccant to keep your mushrooms from touching it.

Arrange your mushrooms on the rack, ensuring they're not too close together, and seal the container.

Wait for a few days, then test to see if they're cracker dry.

Transfer to storage bags (e.g., Ziploc, vacuum-sealed) and place in the freezer.

➤ Reusingreusing The Substrate

After your first flush, the same cakes can be reused up to three times. Dry them out for a few days and repeat Step 5.2 (dunking). But don't roll them in the vermiculite; place them back in the grow chamber and mist and fan as before. When you start to see contaminants (usually around the third reuse), drench the cakes with the mister spray and dispose of them outside in a secure bag.

➤ Making Spore Syringes

Filling your psilocybin spore syringes is about as self-sufficient as it gets.

First, you'll need to take a spore print from a mature mushroom, i.e., one that's been allowed to grow until its cap has opened out and the edges are upturned. You should also notice an accumulation of dark purple deposits around the base. These are the magic mushroom spores.

To collect them, remove the cap with a flame-sterilized scalpel and place it gills down on a sterile paper sheet. Cover with a disinfected glass or jar to protect it from the air and leave for 24 hours. Keep the resulting spore print out of light in an airtight plastic bag.

To load a spore syringe, scrape some of the spore print into a sterile glass of distilled water. You can find this at auto supply stores. Then fill your syringe (which should also be sterilized) and empty it back into the glass several times to evenly distribute the spores. Fill it a final time and place it inside an airtight plastic bag. Leave at room temperature for a few days to allow the spores to hydrate.

You can then keep the syringe in the fridge until you're ready to use it. It should last at least two months.

> **Adaptations And Alternatives**

Numerous modifications have been made to the PF Tek method, both to increase yield and to make things easier. Different species also tend to produce better with different substrates and conditions.

The main alternative to the basic PF Tek is the monotube method, which involves spawning to bulk on coir (coconut fiber extract), manure, straw, or some other fresh and nutritious substrate. Eventually, you may want to experiment with some of these other methods, but the PF Tek is an excellent introduction for now.

Healing Mushrooms

The Healing Mushroom refers to a bunch of small, edible mushrooms. Healing Mushrooms have white stems and red caps and are easily identifiable by their distinctive red glow. They may be found in bunkers, sewers, caves, overgrown areas, or by water bodies. Consuming a Healing Mushroom yields a 100% chance of curing poisoning.

➤ Medicinal Mushrooms - Nature's Perfect Healer

Many of us have heard of and cooked with some of the most common culinary mushrooms: portabella, white button, etc. However, there is a whole class of mushrooms called medicinal mushrooms, and these medicinal healing mushrooms are not meant for flavor-enhancement, though some of them can be used in recipes. They are often taken in tea form, tinctures and extracts, and in capsules as powders.

These medicinal mushrooms include reishi mushroom, Agaricus Mushroom (or Agaricus blazei Mushroom), maitake, shitake, and Coriolus Mushroom. There are many others, but these are some of the most popular. Medicinal mushrooms such as these share much in common with human beings in terms of their chemical and genetic structure.

Many scientists say that mushrooms are closer to human beings genetically than almost any other plant. Given this fact, certain "higher level" mushrooms, often called the "medicinal mushrooms" (NOT magic mushrooms!), can positively heal and impact the body, emotions, mind, and spirit of we humans who consume them.

Reishi mushroom is one of the preeminent healing mushrooms of China (though it is found in other parts of the world as well). Reishi mushroom is sometimes referred to as "the mushroom of immortality" since it can be taken every day as a tonic "herb," and it useful in extending life.

Reishi mushroom is frequently used (in the East) as an immune system stimulant by people with HIV and cancer. Reishi is also purported to help reduce inflammation, help with fatigue, help heal viral issues in the body, and to help calm and relax the spirit, assisting people in meditating and connecting with spirit easier.

Maitake mushroom is another one of the very most powerful medicinal mushrooms in the world. Maitake is used culinarily as well as medicinally and is a potent source of beta-glucan polysaccharides--potent immune system healing chemicals.

Maitake mushroom is used in cancer prevention, as well as helping control diabetes and high cholesterol. Maitake increases the activity of the natural killer cells of the body, helping rid the body of immune system problems.

Just one final note, if you are thinking of checking into the healing properties of medicinal mushrooms: if you take them supplementally alongside a Vitamin C source (preferably a natural plant-based Vitamin C source), you can triple the effectiveness of both the medicinal mushrooms AND triple the effectiveness of the Vitamin C! Powerful information for having the best health ever!

Michael Goldman is a health educator and online writer about super health and super nutrition, primarily through raw and living foods.

➢ The Black And White Medicinal Mushroom Show

Some types of edible mushrooms or fungi like black and white fungus are an essential source of vegetable proteins, minerals, vitamins amino acids, and phytochemicals in a plant-based diet. They have long been in use as medicine and food in Asia though they are only making headway in the West in the last decades.

➢ Wild And Cultivated Mushroom Fungi

Both the black Auricularia and white tremella have a particular affinity for deciduous trees. While it pleases the black Auricularia to grow in wet evergreen forests, the white tremella is more commonly found in temperate forests. The translucent white, fronded, gelatinous sprinklings of tremella on the branches are a beautiful sight, like masses of fresh manna from heaven! However, these two mushroom fungi types can also be commercially cultivated.

➢ Anti -Tumor And Immune Properties

Both Auricularia and Tremella are good sources of polysaccharides, a compound, tested and tried for its anti-tumor and immune-stimulating properties. These medicinal fungi behave like adaptogens in helping your body systems to develop resistance to illness and infighting tiredness.

You are bound to love the black Auricularia in savory dishes of ridge gourds with just a small handful of cellophane noodles, and the tremella is best enjoyed in dessert soups sweetened with jujubes and dried logans. However, all dried fungi must be soaked in water for at least thirty minutes to turn them into globby bits of goodness.

➢ Collagen Properties

Women chasing for beauty may be glad to hear Auricularia and Tremella are high in vegetable collagen. So you can eat yourself beautiful without botox injections or any other cosmetic procedures! There are not that many completely plant- sourced forms of collagen besides the two fungi.

Auricularia and tremella make great food choices as they are affordable, easy to prepare, and delicious to boot. They fight the fat and cholesterol and protect your heart; they battle cancer and give you a new lease of life. Moreover, these healing mushrooms are chock-full of phytochemicals too.

Whatever reservations you may have about eating the two fungi, they will slip away with this -that I practically grew up eating the mushrooms, and there is seldom a day when I don't take Auricularia. My years of eating these foods have certainly repaid me well, and that I have guarded my heart well after all!

From the Foragers at the Verulam Arms, I learn of the delicious sparaxis crispy Mushroom, also known as 'cauliflower of the woods'; I am particularly excited over the fact that it looks like tremella, except that it is much more significant. As for the Auricularia, it is currently grown locally, so it is no wonder I get to eat it every day - well, almost!

➢ Healing Mushrooms To Add To Your Diet

A workout for the immune system is how Tero Isokauppila describes the extraordinary health-promoting benefits of regularly consuming fungi.

Isokauppila, founder and president of Four Sigmatic which makes mushroom powders (dynamite mixed into smoothies), mushroom coffee, and mushroom hot chocolate, to name just a smattering of their prettily packaged, salubrious offerings grew up in Finland, foraging for mushrooms on his family's since-the-1600's farm.

He's contributed to goop as an expert before, helping us understand all of the purported energizing, soporific, immunity-boosting, and beautiful-skin-promoting effects possible with shrooming.

In his new cookbook Healing Mushrooms, Isokauppila focuses on the ten mushrooms most important in helping with day-to-day stress management and sustaining health. He works these into 50 easy, sumptuous recipes. The most foolproof of which, he says, is the Reishi Chocolate Almonds scroll down for the recipe. (Note: the cordyceps-infused cocktail, mushroom bacon, lion's mane latte, and most of the other methods also require no exceptional chef skills.)

Isokauppila's additional labor of love is the newly opened Shroom Room on Abbot Kinney in Venice, CA, a (mushroom) coffeehouse hangout that serves up an assortment of deliciously inventive beverages like adaptogenic lemonade. Shroom Room has the same mission as Four Sigmatic to educate people about the spectacular power of mushrooms.

"People who take mushrooms daily, especially those who mix it up and take a variety of types, some days a little lion's mane, another day more cordyceps seriously benefit as they may see a change in overall well-being," Isokauppila says. This is why our copy of Healing Mushrooms is already dog eared. Below, we asked him to tell more about the powers of mushrooms and how to incorporate them into our diets:

Mushrooms For Cancer

Healing with herbs is becoming more popular as an adjunct therapy, especially in cancer treatment. But what about improving with mushrooms? Herbs are considered plant material. Where do the fungi fit in? From a botanical standpoint, herbs are herbaceous plants. Leaves, roots, and flowers of herbs may be used in herbal medicine. The plant kingdom is comprised of plants.

The fungi kingdom is comprised of mushrooms. Science is taking a closer look at the value of medicinal mushrooms in treating severe medical conditions, including autoimmune disease, nerve disorders, and cancer. This book uncovers some of the mystery surrounding mushrooms and takes a brief look at their use in natural medicine.

Many myths are surrounding the Kingdom Fungi. You may be thinking, "Well, some mushrooms are poisonous." And yes, this is true. Some plants are also poisonous. Mushrooms get a terrible reputation because often cases of mushroom poisoning attract a lot of attention. Most mushrooms are not toxic. You may be thinking, "Will I see visions or hallucinate?"

Many cultures around the world use hallucinogenic mushrooms for healing. But medicinal mushrooms are being researched in laboratories around the world, and practicing physicians, cancer doctors, and alternative medicine practitioners are taking mushrooms seriously and prescribing them for severe medical conditions.

Which mushrooms are medicinal? When searching for mushrooms for healing cancer, look first to the polypores, or shelf fungi. These mushrooms are the oldest from an evolutionary standpoint. Some mycologists (those who study fungi) believe that all mushrooms have evolved from polypores.

Polypores are hard, not soft like gilled fungi. For any mushroom to be digestible, it must first be cooked, heated, or tenderized. This is especially true in the case of polypore mushrooms. They must be boiled first to be bioavailable. Historically polypore mushrooms have been heated and steeped in hot water, strained, and the resulting drink served as a mushroom tea.

Historically, polypore mushrooms were quite valuable to native peoples all around the world. Some hard, shelf mushrooms were used as fuel or spunk to start fires and carried over long distances. These same species were also chopped up and steeped in water for tea. Shamans in cultures on every continent treated serious medical ailments with polypore mushrooms.

Which mushrooms for cancer treatment are polypore mushrooms? The most famous and widely used polypore mushroom is the Reishi mushroom. It is used extensively in Traditional Chinese Medicine, by mainstream Japanese physicians, and throughout Korea, Vietnam, and Eastern cultures. Also known as the Ling Chi, this medicinal Mushroom is available in supplement form over the internet and directly from alternative medical practitioners.

Another essential medicinal polypore mushroom is Grifola frondosa, also known as maitake. Maitake is a soft fleshed polypore with nutritional and medicinal value. It is attracting a lot of attention from pharmaceutical and nutraceutical companies because initial studies show it is quite useful as an anti-tumor medicine, especially in cases of liver and breast cancer. Look for Maitake supplements that address the D-fraction and beta-glucans.

Maitake supplements are widely available over the internet and from natural pharmacies. After dispelling some of the myths around mushrooms and briefly exploring the history of mushrooms as medicine, it makes sense that the mysterious fungi hold healing power.

➢ Mushrooms For Cancer From Around The World

These days, the buzz in alternative and complementary cancer treatments is all about mushrooms. It seems unlikely. Mushrooms can be slimy, poisonous, and downright fungal. But science is taking a close look at medicinal mushrooms and how they fight cancer, tumors, chronic fatigue syndrome, and autoimmune diseases.

More studies are coming out every day, showing that medicinal mushrooms strengthen immunity, fight cancer, and shrink tumors. If you are searching for mushrooms for cancer treatment, this book gives you a brief overview of mushrooms from around the world that have proven efficacy in the fight against cancer.

In the research on medicinal mushrooms, science is proving what Eastern

Medicine has known for centuries. Mushrooms have been used extensively in China and Japan for thousands of years. Doctors of Traditional Chinese Medicine use mushrooms like Reishi, Shiitake, Maitake, and Lion's Mane to promote longevity and keep the body systems healthy and active. Mushrooms help the flow of "chi" throughout the body, increasing energy, removing toxins, and creating an overall feeling of well being.

Science is taking a close look at mushrooms in the fight against cancer. More studies are being performed on the anti-cancer effects of wild mushrooms. In many parts of the world, derivatives of wild mushrooms are being used to treat cancer.

One of the most famous medicinal mushrooms is the Reishi, also known as Ganoderma lucidum. This mushroom is known as "The Mushroom of Immortality" and has been prescribed for at least 3000 years in Eastern Medicine. In 1990 the Japanese government officially listed Reishi mushroom as an adjunct herb for cancer treatment. Reishi is being used with favorable results in cancer research centers around the world.

Reishi is perhaps one of the world's most well-known, widely used medicinal mushrooms. It is available in the complete form, which is a thick, shelf mushroom, and also in tincture, powdered, and pill form. Chaga, also known as Inonotus obliquus, is being used around the world for treating cancer. Historically, the mushroom has been used in Poland and Western Siberia for centuries Russian folklore tells of a fungus that grows on birch trees that is effective in treating a variety of cancers.

Chaga "tea" is an infusion of the mushroom given to cancer patients. The U.S. National Cancer Institute has reports that Chaga has been used to treat cancer successfully. Chaga is proving to be one of the most important medicinal mushrooms for cancer treatment.

One of the most widely available mushrooms for cancer treatment is Tramates Versicolor, also known as the Turkey Tail mushroom. It grows prolifically throughout North America and is used commercially around the world to fight cancer. P.S.K., known as "polysaccharide Kureha" is a derivative of the Turkey Tail mushroom.

In scientific studies, P.S.K.'s anti-tumor activity is enhanced in combination with radiation and chemotherapy. Improved survival rates in patients who take P.S.K. are widespread. In one study, the price of cancer deaths within five years was 21% with P.S.K. and 52% without.

Three famous mushrooms for cancer from around the world include the Reishi, Chaga, and the Turkey Tail. These are all available online and at local health food stores. If you are looking to improve your overall health and wellness, be sure to include medicinal mushrooms in your diet or supplement with teas, tinctures, and pills.

➢ How Do Mushrooms Help Fight Cancer?

The Agaricus blazei mushroom is made up of Beta-(1-3)-D-glucan, Beta-(1-4)-a -D- glucan & Beta - (1-6)-D-glucan. Known as Beta Glucan, these immune-enhancing substances are proven to have potent anti-tumor properties. While they do not directly cause the anti-tumor effect, they do trigger the bodies' anti-tumor response. An anti-tumor white blood cell known as Natural Killer cells (NK cells) is produced by the body making the level of NK cells in the body relatively easy to measure.

When human subjects are given Agari in their diet, a 300% increase of NK cells in the blood is seen within 2-4 days. Natural killer cells are best known for their capacity to kill tumor cells before they become established cancers, but there has also been evidence for their role in controlling infection in the early phases of the immune response by the body.

➤ Mushrooms Can Reduce The Risk Of Colon And Prostate Cancer

What do we know about the humble Mushroom? Well, they have been enjoyed for thousands of years; they are rich in vitamins and minerals, rich in high-quality protein, and are low in fat. They have been used in medicines for ages. In short, mushrooms have been around seemingly forever. But in recent times, studies have shown that certain mushrooms have properties that may well help our body to build a stronger immune system and thus be better able to fight cancer.

The mushrooms mentioned are maitake, shiitake, reishi, Coriolus Versicolor, and

Agaricus blazer Murill. Your greengrocer will be able to help you find these different types of Mushroom. It's the potent compounds found in these mushrooms, which help us in fighting cancer. Mushrooms contain the protein lectin, which prevents cancerous cells from reproducing.

Interestingly, modern scientists have studied mushrooms and concluded they have cancer-inhibiting qualities. It's interesting because mushrooms have formed an essential part in medicine and healing for thousands of years. In Ancient and modern Eastern medicine, the Shiitake, Reishi, and Maitake mushrooms are used for healing. Note that today's scientists put these very same three mushrooms in the list of those which have anti-cancer qualities.

And it's not just cancer which the mushroom helps fight. The standard white button mushrooms contain phytochemicals that are believed to help protect us against breast and prostate cancer. And this same simple mushroom also contains selenium, which studies show helps reduce the risk for colon, lung, stomach, and prostate cancer. It's a tiny plant, but it packs a powerful punch in the fight against cancer.

Some of the statistics which come from scientific studies are quite amazing. A group of 2000 women in China were surveyed and had tabs kept on them over some time. Those women who regularly ate fresh or dried mushrooms were less likely to have breast cancer than those women who didn't eat mushrooms. Those women who ate 10 grams or more of mushrooms every day were 64% less likely to have breast cancer than women who didn't eat mushrooms. Even allowing for the percentage difference on surveys, that figure is remarkable.

Those women who ate 4 grams of mushrooms daily were 47% less likely to develop breast cancer than those women who did not eat mushrooms. This, too, is remarkable. 47%, if you needed reminding, is almost a half. We're not talking about a small percentage here of say 5 or 10%. These are massive changes in the likelihood of contracting breast cancer. And all thanks to the humble mushroom.

Again let's go back to ancient medicine. The ingredient Cordycepin has recently been the basis of a new cancer drug. It came from a wild mushroom and was used for medicinal purposes thousands of years ago. It now seems that modern science is using ancient science to create modern-day cancer drugs.

But remember that the mushroom, eaten regularly, has always been a boon for your health in fighting cancer. As isolated as you may feel right now, you are not alone. A cancer diagnosis is no longer the end of the story. Cancer survivors

➤ Effectiveness Of Mushrooms In Cancer

Mushroom extracts are very effective in preventing and curing cancers. It provides you with anti-cancer properties, and you can safely consume it for a long time without any side effects.

Mushrooms are vegetables that blend very well with soups, stir-fries, and salads. But many are unaware of the fact that actually, it is a type of fungi that grows and feeds upon decomposing plants and trees. Some mushrooms grow on some specific trees and under certain conditions. Chinese use mushrooms in treating colds, pains, and allergies.

Medicinal mushrooms are used as capsules, in tea forms, and as extracts. Reishi mushrooms, Agaricus mushrooms, maitake, and shitake are various medicinal mushrooms. These mushrooms have healing properties, and hence they are used by the medical fraternity.

Reishi mushroom is found in large numbers in China. This can be consumed every day and is a perfect tonic. It is thus termed as the mushroom of immortality. It helps you to keep fit for your entire life. This mushroom is effectively used as an immune system for patients who are suffering from cancer. It helps in reducing fatigue. These mushrooms have anti-inflammatory properties.

It also provides you to be calm and thus helping you to relax. Dried reishi power was viral in ancient China. It demonstrated the anti-cancer activity by destroying cancer cells. This mushroom can also act as a dietary supplement because it exhibits therapeutic properties. These mushrooms can act as an alternative therapy for breast cancer and prostate cancer.

Thus, reishi mushrooms provide cures for multiple diseases. It helps in maintaining the body's independent balance. These mushrooms can be consumed for a long time and that too without any side effects. It also helps in maintaining the natural resistance of the body.

Mushrooms have low calories and have 80-90% water. Maitake mushrooms have low molecular polysaccharides, which helps in increasing the immunity of people. It energizes the immune system and helps in attacking the pathogenic system.

Maitake mushrooms reactivate immune-competent cells, thus enhancing the functions of macrophages and T cells.

In this way, it helps in efficiently providing a solution as an anti-cancer solution. It contains beta-glucan, which is used very effectively in anti-cancer therapy. These glucans produce T cells and NK cells, which safeguards you against cancer. The Shitake mushroom is beneficial in providing immunity to your system.

The consumers are required to be very cautious about the false advertising of mushroom extracts for preventing and curing cancer. This is because many beta- glucan products that are available are not 100%

pure. It may so happen that these mushrooms may contain only 1% beta-glucan. Thus you are required to read on the labels about its purity before you purchase it.

Ganoderma Mushroom

Ganoderma is an herb that comes from the Reishi mushroom. Many herbs are taken for different medicinal purposes, but Ganoderma surpasses them all. Thus named the "King of Herbs," it is packed with powerful antioxidants and offers multiple health benefits.

In the US, this herb has not reached a level of popularity where it could be called well-known. If someone is asked if they have heard of Ganoderma in this region of the world, it would not be surprising for the response to be, "gano-what?" Even spell-check - a tool trusted and utilized by people in all walks of life - does not contain the word Ganoderma. Ganoderma does have a proven track record, though; it has been used in traditional Asian medicine for over 4,000 years.

➢ Health Benefits Of Ganoderma Lucidum

Ganoderma has been known for centuries for alkalizing and oxygenating the body to establish the foundation for a lifetime of good health, removing the basis for osteoporosis, arthritis, adult-onset diabetes, heart disease, and many other degenerative conditions including cancer. No virus can survive in a super oxygenated environment. When your body has sufficient oxygen, it thrives.

Ganoderma is a super herb which rapidly oxygenates your body and automatically adjusts your PH to a healthy balance at the cellular level.

Weight Loss: Oxygen is a necessary component for burning fuel. The more you oxygenate the body, the higher your metabolism. That means weight loss.

Ganoderma is backed up with evidenced-based documented research in reducing cholesterol levels and easing allergy-related inflammation of the airway. Often used as an immune stimulant by people with cancer (as well as HIV), Ganoderma has been shown to strengthen immunity as well as combat cancer-cell proliferation

Other Health Benefits Of Ganoderma Lucidum Include The Following:

- Helps balance blood sugar and pancreatic functions
- Helps protect against skin cell degeneration
- Helps improve skin texture
- Helps reduce the appearance of aging
- Helps eliminate toxins from the body
- Helps increase metabolism
- Helps fight and inhibit free radicals
- Helps unclog arteries and supports liver function
- Helps improve sexual services
- Helps reduce fatigue and improve sleep
- Help improve the digestive system
- Relieves sinuses congestion and other respiratory problems
- Helps rejuvenate body tissues and cells

- Helps increase brain power and clarity
- Helps provide energy and vigor
- Helps rejuvenate and oxygenate the body
- Helps strengthens the immune system
- Helps support the regulation of blood pressure
- Helps balance cholesterol
- Effectively aids in the healing of skin wounds, scrapes, psoriasis, mouth ulcers, external bleeding, bug bites, and stings.

➤ The Different Forms Of Ganoderma

The contents of Ganoderma vary at different stages of the mushroom's growth. As a result, Ganoderma is harvested at its different stages of growth for use in various products. The three forms are Ganoderma mycelium, Ganoderma lucidum, and Ganoderma spores.

➤ Ganoderma Mycelium

Ganoderma mycelium is a product of the reishi mushroom when it is in root form. At this stage of its growth, it is only 18-21 days old. In this stage, it contains a high concentration of organic germanium and polysaccharide. Ganoderma mycelium has positive effects on the human brain and is also an immune system enhancer.

➢ GANODERMA Lucid

Ganoderma lucidum is a product of reishi mushroom when in its mature state. It reaches this state after 90 days of growth. This is a long list of health benefits associated with the mushroom in this form. To name a few, Ganoderma lucidum is an anti-allergen, antioxidant, anti-inflammatory. It also balances the pH levels in the body and has a regenerating effect on the skin.

➢ Ganoderma Spores

Ganoderma spores are like the seed of the mushroom. Once the spore is cracked, its health benefits can be reaped. Organo Gold has a patented technology that allows them to break the spore to yield the maximum benefit. Ganoderma in this form is 17 times more potent than Ganoderma lucidum and is rich in polysaccharides, triterpenes, Organic germanium, and selenium.

The Ganoderma Coffee And Reishi Mushroom Connection

The red reishi mushroom and Ganoderma lucidum are the same. They are simply two different words that are used for the same mushroom, the one that has been used in Chinese medicine for hundreds of years. In

Ganoderma Coffee, the red reishi mushroom is combined with robust Columbia coffee or other coffee beans to make a very rich-tasting cup of coffee. Moreover, you will not be able to taste the reishi at all because it is mixed with the coffee, but you will reap the countless health benefits from drinking the beverage.

The most significant benefit of the reishi mushroom is the fantastic way it helps boost the immune system, which is vital for people suffering from chronic illnesses. If you're prone to colds and allergies, this coffee will work miracles in your life and overall well-being. The benefits are often noticed within the first week of drinking coffee daily. You will begin to feel better and start seeing you are not getting sick as much as you did in the past.

Whether you prefer to call it Ganoderma Coffee or red reishi coffee, the benefits you will receive are the same. The red reishi mushroom has been used in primitive cultures for thousands of years, long before modern medicine was on the scene. Only recently has the world gotten wind of what early societies have always known, that this mushroom can work wonders in keeping people healthier.

Often it used in Chinese medicine as part of an overall wellness regime. This coffee is an excellent addition to anyone's diet no matter regardless of your health, whether good or not. There are currently no known side effects to Ganoderma, and it has no addictive properties. It is also touted as one of the safest and most regarded herbs on the market today, and many people are discovering every day. Not too many herbal supplements, and more importantly, gourmet coffees can be thought of as a way to improve overall health.

Ganoderma Coffee is often used as an alternative to traditional medicine to treat disease, or it can be used as a complementary treatment to conventional medicine. If you are allergic to prescription drugs that your doctor prescribes to you or you, do not tolerate the drugs well, speak to your doctor about the possibility of replacing it with red reishi coffee beverage. They may or may not agree, but it's important to let at least your medical provider know what you are indenting on doing.

This coffee offers a beautiful part in helping to prevent cancer, heart attacks, high cholesterol, and elevated blood pressure. Ganoderma Coffee acts as a stress reliever as well as a mood stabilizer, thus alleviating anxiety and depression. Although the red reishi mushroom has been around for thousands of years, it has only been in recent years that Western society has begun to take advantage of this beautiful ancient healing mushroom.

➤ What's So Special About Ganoderma Mushroom?

Ganoderma mushroom is from the Mycetes kingdom. These are generally short and tiny fungi that are unable to manufacture their nutrients through photosynthesis like green plants.

They can use either Earth breed or lead a parasitic life by breaking down and surviving on nutrients of other plants and animals. However, due to the increasing worldwide popularity of Ganoderma mushroom as a medicinal herb, this fungus is now mass-produced with a scientific cultivation method.

Within the kingdom of Mycetes, Ascomycetes are the lower grade, while Basidiomycetes such as Ganoderma Lucidum and mushroom are of a superior grade. The Basidiomycetes have a very close relationship with humanity. Many are edible and have excellent healing properties. These fungi naturally become the more precious medicinal herb in the Basidiomycetes kingdom.

Generally, mushrooms' families can improve the immune system by activating Macrophage and Natural Killer (NK) immunity cells. A macrophage is one of our essential first-line defense against viruses, bacteria, and foreign substances. NK cells are responsible for killing cancer cells. Some can even improve allergic conditions such as skin allergy, nose allergy, and asthma.

Some of the beneficial medicinal mushrooms are:

- Golden needle mushroom (Flammulina velutipes)
- Shitake (Lentinus edodes)
- Maitake (Grifola frondosa)
- Agaricus Blazei (Agaricus blazei Murrill)
- Reishi mushroom (Ganoderma lucidum)

All these fungi can prevent high blood pressure, diabetes, hardening of the blood vessel, and even cancer. But of all the fungi, Ganoderma Lucidum significantly stands out in its medicinal values.

However, you may ask. Among these mushrooms, it is proven that Agaricus Blazei Murill (ABM) has the highest concentration of beta-glucan polysaccharides. So, does Agaricus Blazei Murill mushroom has the best anti-cancer effect? That's right. ABM is the best mushroom for the anti-cancer property. But it is comparing AMB with Ganoderma Lucidum, which one is better?

Ganoderma Lucidum does not only contain beta-glucan for anti-cancer, but it also contains Organic Germanium. This precious mineral has been proven to have a strong anti-cancer effect too. Having both beta-glucan and organic germanium together greatly enhances Ganoderma mushroom's ability to fight cancer.

Besides, Ganoderma Lucidum contains Triterpenes to help cleanse our blood vessels from cholesterol and fatty substances. Ganoderma mushroom even has

Superoxide Dismutase (SOD). This enzyme functions as our primary defense line against free radicals in our bodies. SOD in Ganoderma Lucidum can help to protect against attack from harmful free radicals. This makes Ganoderma mushroom one of the best fungi for anti-free radical and anti-aging.

In a Chinese medical encyclopedia, "Ben Cao Gang Mu" (Materia Medica -

Detailed Outline of Medicinal Herb), an ancient medical practitioner Lee Shi Zhen in the Ming Dynasty classified Ganoderma Lucidum as a "superior herb."

He grouped Ganoderma mushroom into six categories by its color and form, namely Green, Purple, Red, Yellow, White, and Black Ganoderma. All six types have their medicinal values and healing effects. According to the record, "continuous consumption of Ganoderma mushroom makes the body light and young, lengthening life and making one like an immortal who never dies."

Later, modern medical researches on Ganoderma Lucidum proved the findings of those ancient medical texts. That is why, for over 2000 years, Ganoderma mushroom is highly praised in China. The Chinese call it the auspicious herb, which also influenced their literacy and arts.

Acupuncture & Mushroom Nutrition - Secret To Longevity

Some of the earliest written Chinese medical texts dating back 3000 years and more place certain rare mushroom species amongst the most potent types of healing herbs available in the world today. Respected for their proven ability to promote optimum health, longevity, and wisdom without adverse side effects, they were for many centuries available only to the wealthy merchant or ruling classes of ancient China due to their rarity and prohibitive cost.

> ➤ **Secret To Immortality**

You may know of Mushroom species such as Reishi, Cordyceps, or Shitake? As mentioned above, these were once reserved only for the wealthy classes or royalty. Known as "The Emperor's Secret to health," they were available only to him and his family due to their rarity and expense. With them lay the keys to health, vitality, longevity, and some even believed immortality.

Specific strains of Mushrooms have Immune-regulating properties, which can result in many health benefits. Recent research has shown Mushroom Nutrition to be beneficial in **the following areas:**

- Mental / Emotional
- Improve Concentration Lift Depression Moderate Anxiety
- Fertility
- Promotes Fertility Regulates Menstruation Alleviates PMT Reduces Period Pain
- Sports Performance
- Increased Energy Increased Endurance Injury Recovery Increased Blood Oxygen Levels
- Circulatory System
- Hypertension Hypotension Arteriosclerosis Cholesterol Anemia Elevation Sickness
- Digestive System
- Gastroenteritis Liver Necrosis Obesity Diabetes Constipation Gallstones Appetite

Are Chaga Mushrooms Anti-Aging And Healing?

Do you like mushrooms? Do you eat them often? Do you know what kinds of mushrooms you eat and what nutrients they contain?

Many of us have grown up disliking and even fearing mushrooms. I remember being told that some types of wild mushrooms are dangerous and can make you deathly ill. While that is still true, what I have since discovered is that some mushrooms are truly medicinal. Some mushrooms have so much healing potential that it is indeed a shame that more people don't realize this and take advantage of this marvelous healing food.

We have all enjoyed casseroles and meat dishes flavored with mushrooms and mushroom sauces. Some of us love the taste of stuffed mushrooms or our favorite Asian and Oriental dishes that include mushrooms as a staple ingredient. As a food, especially when eaten along with other foods that may camouflage the taste and appearance, most of us have enjoyed mushrooms on occasion.

However, most of us do not understand the medicinal value and true healing potential of certain mushrooms, such as Reishi and Chaga. One common ingredient in medicinal mushrooms is polysaccharides, a type of complex carbohydrate that activates cancer-fighting blood cells, macrophages, and T- lymphocytes, stimulates interferon, and improves the immune function of the cells. While polysaccharides stimulate immune function in the body, they appear to have no toxic effects the way, so many pharmaceutical drugs do.

Medicinal mushrooms help to normalize the body's defense system, thereby assisting the body to resist and fight off diseases. Chaga mushroom, which grows in one of the coldest regions of the world, Siberia, has been known to suck all the nutrients out of the birch tree and then the tree just naturally dies. The Chaga mushroom retains all of those healing nutrients, which can then be transported into the human body, often in the form of a tea or an extract. Some extracts are made with alcohol, which depletes some of the powerful antioxidants.

If you have ever had a live blood cell analysis taken, you will discover what is lurking in your blood. I found some free radicals, the kind that leads to cancer, and some candida and bacteria in my blood. Immediately, I took some of my

Chaga extract and my CoEnzyme Q10. One month later, when I returned for my second live blood cell analysis, the free radicals were gone entirely!

The bacteria were also gone. I no longer have any doubts about the power of antioxidants and the other powerful immune-enhancing substances found in medicinal mushrooms, especially Chaga. Mushrooms, particularly Chaga mushrooms, may hold the key to strengthening your immune system. With all the toxins in our food and even in the air and our environment, we owe it to ourselves to do whatever we can to keep our immune system in optimal health, especially as we age.

Top Advantages Of Reishi Mushrooms, Also Known As Ganoderma

Much research has been done on the Reishi mushrooms and the benefits it has on cancer treatment. This was done through in-vitro and human studies experiments. Next to ginseng, Reishi mushrooms are rated at the top among Chinese traditional medicinal herbs. In China, it is known as LingChi or Lingzhi, while in Japan, it is recognized as Reishi. In the west, however, it is widely known Ganoderma Lucidum. It has been used for over 2,000 years by the Chinese as part of their medicinal treatments.

In Asia, it is commonly and widely used in treatment for cancer as well as immune-stimulating prescriptions. Many medicinal mushrooms research has been done on the Reishi mushrooms and its benefits that it is regarded as one of the most respected herbal medicine in Asia. The research was done in Japan, Europe, and parts of America shows that one common element of all the studies is that

Reishi mushrooms have the ability to affect the immune system most unusually and positively. Because of this, it benefits a large number of illnesses, including various forms of cancer.

Other than improving the immune system, researchers also found that it has anti- inflammatory, anti-allergic, anti-viral, anti-bacterial, and antioxidant properties.

Here Are The Top Advantages Of Reishi Or Ganoderma Lucidum Mushrooms:

- Reduce cholesterol and enhancing metabolism;
- Purifying blood hence detoxify the body resulting in revitalized cells;
- Enhancing liver functions and remove free radical;
- Dissolve thrombosis thus prevent stroke/heart attack;
- Normalize immunity (polysaccharide + triterpenoids);
- Improving the elasticity of the blood vessel;
- Builds up immunity resulting in better antibody

Studies also show that Reishi has the ability to inhibit cancer cells and produce new capillaries and thus making it a very potent anti-tumor. In has also been known to reduce the side effects of radio and chemotherapy and stimulate the appetite of a cancer patient.

So, who can benefit from Reishi? People with diabetes, hypertension/hypotension, asthma, kidney, liver, brain, intestine problems, cancer, gastric, gout, cyst, fibroid, heart, and many more. It is even beneficial for people with no illnesses as it strengthens the body's immune system. The simple reason that it is so helpful is because of its fantastic and positive natural way that it affects the immune system.

For many generations, Reishi mushrooms have been used in Japan and China as medicine for hypertension, arthritis, and liver disorders. Reishi mushrooms have no side effects, and because of this, it could be taken with doctors' medication. It is even safe for children and pregnant women. It can be found mostly in tablet or capsule form, although be sure to check that it is of excellent and superior Reishi to get the full benefits.

Grow Mushrooms For Food And Other Reasons

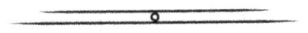

As a fungi food variety, the mushroom is produced above the ground cultivation. These come in different varieties, and you need to know the type you want to farm. There are wild and edible mushrooms. Edible mushrooms are nutritious and provide a good source of vitamins and mineral sources to the diet. They taste like meat, and vegetarians and vegans will find this delicious.

As you choose the types of mushrooms you want to grow, consider the various types, including; white, crimini, and Portobello varieties. You could also find options of the oyster, maitake, and shiitake mushrooms. Edible mushrooms will provide a great meal. You can blend mushrooms with other foods to make tasty meals.

> **Factors To Consider When Farming Mushrooms:**

Once you have collected the edible mushroom variety, you need to consider some factors under which they can grow. This includes the weather, moistures, soil, and fertilizers. It Is Appropriate To Find:

- Type of Mushrooms to farm and the seed to use (edible, medicinal and fiber mushrooms)
- Available land and soil type
- The temperature under which they need to grow
- Farming Procedure for the Mushrooms including cultivation, weeding and harvesting process

You can produce mushrooms for commercial or local consumption. If you are farming the mushrooms for your household, the amount produced might not be much. However, farming mushrooms for business requires that you look for the market since these are perishable food produce. As a producer for the business market, you might want to consider the dried mushroom options, which will keep your fruit for longer.

➢ Reasons For Mushroom Farming:

In essence, mushroom farming could be for the following reasons:

- For business
- Home consumption
- Research and Medical Use

➢ How To Plant Mushrooms - Grow Your Own Oyster Mushrooms Indoors

Oyster mushrooms are one of the most natural varieties of mushrooms to grow, and knowing how to plant mushrooms can bring you an almost unlimited supply of mushrooms at your dinner table. Although oyster mushrooms grow in woods, there are other growing media that you can use in raising them. Consider straw and sawdust. They are easier to gather than logs.

Oyster mushroom resembles oysters, and they have a rich culinary and medicinal history to boast. Chinese medicine, from some three thousand years ago, uses oyster mushrooms as a tonic to enhance the immune system. It has ergothioneine, which is an exceptional antioxidant that can protect the cell. Even if oyster mushrooms are cooked, the antioxidant level remains the same. The mushrooms have been proven to possess anti-bacterial properties as well. Oyster mushrooms have significant levels of potassium, iron, zinc, vitamin C, calcium, niacin, phosphorus, vitamins B1 and B2, and folic acid. The study revealed that eating oyster mushrooms contributes to suggested dietary requirements.

Commercially prepared mushrooms have pesticides and other chemicals in them to make them presentable and their shelf life longer. Although fungi can contribute a lot in making you healthy, the presence of harmful chemicals in them might make your life shorter. The solution? Learn how to plant mushrooms and enjoy its many incredible benefits.

➤ Preparations For Your Quest On How To Plant Mushrooms

For this project, you will need two small cardboard boxes or milk cartons for sawdust to fill them in; two cups coffee grounds or whole-grain flour; spawn of oyster mushrooms. If sawdust is not available or if you find it hard to gather sawdust, then you can always use straw as a substitute (although sawdust is much better).

You can begin with a kit if you want, but if you're going to start from scratch, then. Oyster mushrooms can give you a great margin to succeed in your endeavor over other mushroom varieties. Oyster mushrooms have dozens of types to choose from, and you can consult your supplier for the best variety that is appropriate for your location. Most oyster mushrooms grow in places where the temperature ranges from 55 to 65 degrees Fahrenheit.

➢ **The Steps You Need To Follow In Learning How To Plant Mushrooms.**

The steps to follow in how to plant mushrooms are not complicated; in fact, they are easy to understand and follow. It does not require you to be a genius to grow some mushrooms.

You need to cut the boxes that you are going to use to even height or the same size. On the sides of the two plates or cartons, punch several holes (small in dimensions but not as small as a pin).

If you opt to use sawdust that is pre-inoculated with spawn, then don't sterilize the sawdust because it will kill the spawn. If you are using fresh sawdust, then you might want to clean it first. You can steam, boil, or microwave the sawdust. You can cook or simmer the sawdust for a few minutes, and after sterilizing, you can turn off the heat and keep it covered. Let it cool at room temperature before proceeding to the next step.

If you opt to microwave, then you need to get a microwave-safe bowl and put the sawdust in together with the flour or coffee grounds. Fill it with enough water until the mixture looks like a wet sponge. When the water begins to boil, it will kill the organisms that you want to eliminate. You might need to repeat the procedure in the microwave to finish all of your sawdust.

Use non-chlorinated water to wet the sawdust. Make sure that it is thoroughly damp. Carefully blend in your spores. Firmly pack the wet sawdust into the boxes or cartons and leave them in a basement, garage, dark cabinet, locker, or cellar. You can wrap plastic underneath the container and cover them with plastic with some cooking oil sprayed onto them to trap insects if there are any.

Keep the sawdust damp with non-chlorinated water, and in a few months, you will get to enjoy the fruit of your work. When harvesting, make sure to twist the mushrooms gently to avoid breaking the stem.

Learning how to plant mushrooms can be a fun family activity as well that will benefit all in the long run.

➢ The Easiest Way To Grow Mushrooms: Cake Techniques

Most newbies begin growing mushrooms by using what we call "cakes." Cakes are very easy to grow mushrooms from and are not hard to make. You can also purchase pre-sterilized cake kits from vendors if you don't feel up to the task of do-it-yourself. Below we will discuss the different kinds of cakes:

Brown rice flour cakes (B.R.F.): are the most common form of cakes that are used by the novice. Brown rice flour contains most of the nutrients that most mushrooms require and are very easy to make at home in your kitchen.

Wild bird seed flour Cakes (W.B.S. cakes): are made of conventional wild bird seed that you can buy at most grocery or hardware stores. These types of cakes are used less than brown rice flour cakes but work very well for most mushrooms.

Wood Cakes: are used for decomposing wood mushrooms, such as Reishi, Shiitake, Maitake, etc., and are made of supplemented hardwood sawdust, or supplemented hardwood mulch.

➤ The Practical Way To Grow Mushrooms: Bulk Techniques

While cakes are great for the beginner, most novice cultivators move on to intermediate cultivation techniques very quickly after having success with cakes.

The reasons are mostly because the size and yields of the mushrooms are increased substantially, and the cost of cultivation also decreases. However, this is not generally a method used by the beginner because there are more steps to success, which increases the chances of failure.

➢ Bulk Techniques Are Used With Three Different Container Methods:

- 6-12 quart clear plastic shoeboxes
- 60-120 quart clear plastic storage bins

Specially made mushroom growing bags that have a small filter patch glue to them that allows for necessary gas exchange. The mushrooms are produced inside of the clear bag.

➢ Sterile Technique: The Key To Success

Observing the sterile technique is very important if a cultivator wants to have success growing mushrooms. The substrates that we make must be purified or sterilized to kill off any competitor spores or bacteria, which will provide an environment that is favorable for the species that is being cultivated to flourish. This environment must be maintained to ensure that there is no contamination, which will end your cultivation attempt immediately.

It is important to inoculate your substrates in a glove box (which is a sealed container that has gloves attached to holes in the front, like a box at NASA for inspecting moon rocks), or under the sterile airflow of a laminar flow hood. Flow hoods are very expensive, so most people who cultivate at home use a glove box, as one can be made for under $50.00.

➤ Natural Fibers

Mushrooms have a great benefit to the human race and can be produced for food and other uses. Used as medicine, their extract combats numerous diseases, including tumors, and it improves the immune system. To create them for natural fibers, you need a large-scale production, which will require colored mushrooms and fungi types for durable fibers.

Mushroom farming is an affordable activity that will require no money. Since mushroom is fungi, their production multiples to give you much more than expected, to benefit more from the farming activity, you need to research widely on the different types of the fungi and their benefit to humanity. This might convince you to farm mushrooms for much more than food uses.

Different mushrooms require different farming approaches, and the best way is to research their varieties. You can get tips from experienced farmers. You will find mushroom farmers from all over the world with tips on how to handle the various types of produce. As you start on a small scale, you will give yourself time to grow into a large-scale farmer if you are business-oriented. The main reason behind mushroom farming is the benefit it has to you. If you produce for business, take a business approach.

Mad About Mushrooms?

Mushrooms are one of the great world delights - at least for some people. There are those who don't like the thought of eating something classified as a fungus! For all of you who love mushrooms, here's some great information about the various types and a few suggestions for their use.

Five Great Types Of Edible Mushrooms

While there are hundreds of different types of mushrooms that are edible, there are hundreds more that aren't. Some are very poisonous, and others will make you feel as though you are dying! If you are not familiar with differences, then it's best to get your mushrooms from your local grocery store!

There are three types of mushrooms commonly found in the produce aisle. These are white buttons, portabellas, and shiitake. These mushrooms each have a very distinctive flavor when eaten raw, and you won't have to worry about these being poisonous. When cooked, mushrooms tend to take on the character of other ingredients, but add a beautiful earthen tone, as well.

Another type of mushroom is the coveted truffle. In 2007, a 1.5 kg truffle sold at auction for $330,000, entirely out of most kitchen budgets! Most truffles are worth about $1000 to $5000 per pound. This particular mushroom was shaped funny and is why it brought so much. Truffles have a very earthy flavor, and each kind of truffle has its distinct taste and fragrance. It's instead an acquired taste, but if you want to know what this critical mushroom can do to an ordinary dish, pick up a bottle of truffle oil. It will bring even the most boring dish to new heights.

Finally, another type of mushroom that is very popular but not as expensive is the chanterelle. This mushroom has a flavor and aroma that's quite distinctive. It's been described as tasting more like a flower than a fungus, and its fragrance has a distinct apricot essence. This one is a little more difficult to find in stores, but if you have a world market type store in your city, you may have some luck.

Cooking with Mushrooms

Mushrooms are exceptionally versatile when it comes to cooking. White truffles are usually used in tiny amounts; the other types are used in varying quantities. Mushrooms are a great compliment to many foods but can be used as the main ingredient. Stuffed mushrooms with cream cheese and crab meat are a staple for many hors d'oeuvre tables, and fried mushrooms are found in many a "greasy spoon." However, there are many other uses for mushrooms in the kitchen.

Some of the best recipes involve sautéing the mushrooms until they are a darker color. Add in some garlic, onions, and red wine, and you've got the start of a great side dish or a topping for a fantastic steak. Mushrooms are used in many Italian recipes, such as in pasta sauces. Mushrooms are also a great compliment to egg dishes, such as omelets and quiches.

➢ Preparation and Storage

Before using fresh mushrooms, you should rinse them off quickly in cold water, but don't soak them. They tend to become water-logged rather quickly. Most mushrooms can be stored in your refrigerator, but not in the crisper drawer. If you buy loose fresh mushrooms, it's best to place them in a paper bag, as this will help absorb some of the moisture.

It will help the mushrooms stay fresh longer. If you notice that one mushroom is looking rather unfortunate, take it out immediately, as the rest will spoil quickly. Mushrooms are a fantastic addition to many recipes, and you can experiment with the various types, flavors, and aromas to find what suits your tastes best.

Learning How to Farm Mushrooms

Learning how to farm mushrooms can be an exhilarating experience if you are trying to learn how to get into the farm business. Mushrooms are the easiest to start with and are used in a lot of food dishes worldwide to add flavor and diversity. If you are ready to start learning how to farm mushrooms, this is an excellent place to start. By following these few tips, you will be on your way to learning how to cultivate mushrooms and selling them in no time.

- There are hundreds of different types of fungi that you can grow. A lot of these mushrooms are not edible and should be avoided. You don't want to kill or severely injure a person, let alone yourself. You need to research the types of mushrooms that you may be interested in growing, learn how to tell edible mushrooms from bad, and how they develop before you get started trying to learn how to farm mushrooms.
- Mushrooms can be placed anywhere in an open field to be grown, depending on the type that you choose, but you need to find a space

where you can focus solely on learning how to farm your mushrooms. This is great if you have an empty backyard or just an empty plot of land that you are not using for anything. Mushrooms tend to grow in groups, and you will learn this as you are learning to farm mushrooms, and can sprout up anywhere. You want to have an area that you can control your mushroom growing in.

- Do a lot of research on the types of mushrooms that people are looking to purchase. The models that are in demand are the ones you need to grow. You can also add in a few others that you think are different and may be in order soon, if you are business savvy, or you can choose the type of mushrooms that you enjoy eating and grow them so you can supply your household. Either way, you need to know what mushrooms are being used in the world today. When you become the mushrooms that are in high demand, this raises your chances of making a profit because you almost know for sure that the mushrooms you are growing and farming are going to sell. This is not guaranteed, but it is not a complete and total loss either.

- The final step is to build a staff that is going to help you take care of your new mushroom farm. When you are first learning how to farm mushrooms on your own, you may want to do it by yourself. If you are trying to learn how to cultivate mushrooms to break into the farming business, the staff is needed to help you take care of the farm and help you with the financial aspects of having a mushroom farm also. If you are going to have a hugely successful farm, you are going to need all the help you can get. Since you are just now learning how to farm mushrooms and don't know how your budget is going to be looking at the start, you may find some friends or family that would like to volunteer to help get you started. The extra help is always welcomed.

Current Research Fom All Over The World

Joint research undertaken by Zhang from The University of Western Australia and Zhejiang University in China found that eating mushrooms and drinking green tea may protect against breast cancer. Zhang reported that breast cancer was the most common type of cancer among women worldwide and that its rate was increasing in both developed and developing countries. Interestingly, the incidence of cancer in China was four or five times lower than in developed countries.

The study hoped to show if this could be due to the use of dried and fresh mushrooms and green leaf tea in the traditional Chinese diet. Mushrooms, mushroom extracts, and green tea had shown anti-carcinogenic properties, which were thought to stimulate immune responsiveness against breast cancer.

The consumption of mushrooms and green tea by 2,000 women, aged from 20 to 87 in relatively affluent southeast China, was monitored. Half of the women were healthy, and the others had confirmed breast cancer. On the interview, it was found that fresh white button mushrooms, Agaricus sports, and fragrant dried mushrooms, Lentinula edodes, were the most commonly eaten species of mushrooms. Some of the women in the study consumed neither mushrooms nor green tea, while others enjoyed both up to three times a day.

The results of the study showed that the combination of dietary intake of mushrooms and green tea decreased breast cancer risk with an additional reduced effect on the malignancy of cancer. Zhang concluded that, if confirmed consistently in other research, this inexpensive dietary intervention may have potential implications for protection against breast cancer development.

Dr. W. J. Sinden from the University of Pennsylvania and Dr. E. D. Lambert from Lambert Laboratories were the first to present their research results on the medicinal compounds of Agaricus blazei. They attracted the attention of the medical community to this mushroom. Former President Ronald Reagan used this mushroom to fight his skin cancer, which helped publicize Agaricus blazei.

Two Mushrooms Combine For Healthy Living

The Piedade mushroom, found in the rainforests of Brazil, is well known internationally for its healing properties. In particular, the people of the Piedade region who consumed this mushroom were reported to have enjoyed unprecedented health and longevity, many living disease-free well into their 100s. Following several clinical trials, the Piedade mushroom and the Agaricus blazei mushroom, cultivated in the mountainous region of California, were combined to form a super-hybrid and potent mushroom liquid.

Using a 10 stage extraction technology that captures every nutritious element, and combined with Japanese Sasa Bamboo, a powerful antioxidant, this product is considered a powerhouse of nutrients critical to maintaining and sustaining a healthy and active lifestyle. We no longer have to go to the rainforest in

Brazil or climb the mountains in California to find this pure gold.

➢ The Key Product Benefits

Every human is susceptible to aging, environmental contaminants, chemicals in food and water, disease, and the stresses of a fast-paced lifestyle. Furthermore, we could all greatly benefit from effectively enhancing our immune systems. This combined mushroom product promotes health and overall well being. It increases natural killer cell activity, boosts energy, and generally protects the body. It may also lower cholesterol, control blood pressure, and ease arteriosclerosis.

Mushrooms are a nutritional dietary food supplement with an active ingredient proven by research to be an effective enhancing agent to the immune system.

Coupled with your daily intake of activated liquid zeolite, a naturally formed mineral which strips the body of heavy metals and toxins, taking this mushroom product may firmly place you on the pathway to improved health by fighting severe health challenges. Like the people of the Piedade region, you may go on to live a healthy and happy life, enjoying the magic of mushrooms.

➢ Nutrition And Health Benefits Of Mushrooms

Mushrooms are a food which contains several health benefits to our body. The nutritional value of a mushroom includes being low in calories and high in vegetable proteins, iron, fiber, zinc, essential amino acids, vitamins, and minerals. Since the olden history, the Chinese have been using a mushroom due to their rich health content. To the Chinese, the nutrition health benefits of mushrooms include promoting vitality and good health.

Recent scientific studies have confirmed the health benefits of mushrooms. These studies have shown that mushroom strengthens our body and improves our immune system by maintaining physiological homeostasis. The nutritional value of mushrooms differs from the type of mushrooms. However, almost all mushroom brings great health benefits to the human body, and here is some mushroom to discuss.

➢ Shiitake Mushrooms Nutrition

Shiitake has been known as the "Elixir of Life." Shiitake mushroom has been declared as an anti-cancer food by the FDA of Japan. Shiitake mushrooms nutrition includes containing Lentinan. Lentinan is known to have some effect on bowel, stomach, liver, and lung cancer. It boosts the production of T lymphocytes and other natural killer cells and reduces the adverse health effects of AIDS.

Shiitake mushrooms are rich in several other antioxidants such as uric acid, selenium, and vitamin A, C, D, and E. Shiitake mushrooms have been found to lower blood pressure for people with hypertension. In additional to the above nutritional benefits, the precious nutrition value of Shiitake mushroom is known to reduce serum cholesterol levels and increase libido. It stimulates the production of Interferon, which has anti-viral effects. In some studies, it has proven to be effective against Hepatitis.

➤ Agaricus Mushroom Health Benefits

Agaricus mushroom is consumed mushroom in many countries. It is usually regarded as a health food for its medicinal properties. Agaricus is also known as "God's Mushroom" due to its curative health benefits to a wide range of health disorders.

People have consumed it to cure numerous diseases and body disorders relating to the immune system, the heart, and the digestion system. Other Agaricus mushroom health benefits include weight management, controlling diabetes, chronic and acute allergies. Other curative effects include cataracts, stress, and chronic fatigue.

➢ Health Benefits Of Maitake Mushroom

Maitake has other names such as "Dancing Mushroom." It is known for its taste and health benefits after consuming it. In Japan, Maitake Mushroom is also called the "King of Mushroom."

In Japan and China, Maitake Mushrooms have been eaten for the past 3000 years. Back in history, the Maitake is traded as an alternate currency in Japan, and it's is said to worth as much as it's weight in silver! Maitake is used as a tonic and food to help to promote wellness and vitality for the Japanese.

Since history, the consumption of Maitake mushroom was believed to lower high blood pressure and prevent cancer. In the past years, scientists have been experimenting with Maitake mushroom to confirm it's health benefits. Laboratory scientific studies have indicated that extracts of the Maitake Mushroom can control the growth of cancerous tumors and boost the immune system of almost all the cancerous mice used in the experiment.

➢ Nutrition Value Of Cordyceps Mushroom

Cordyceps mushroom strengthens the immune system's ability to fight against viral and bacterial infection. Scientific studies have shown that Cordyceps is useful for the treatment of high cholesterol, impotence, lung cancer, and kidney failure.

Consumption of cordyceps mushroom causes the human muscle to relax. This is a great health benefit that helps treat coughs, asthma, and other bronchial conditions as it smoothens the muscles.

➤ Reishi Mushroom's Nutritional Value Information

In the past, the Reishi mushroom is a royalty food that only the imperial family consume. Reishi mushroom is fondly known as "Ling Zhi" by the Chinese. Studies in the past 30 years have shown that by consuming reishi mushrooms, health benefits such as treatment of common ailments and conditions are reaped.

Recent studies have shown that Reishi has a lot of health benefits and nutritional effects: Antioxidant, Lowers blood pressure, protects the liver, and detoxify it.

What reishi mushroom does is to bring the body's elements back to a natural state, thus enabling all the body organs to regain their functionality again.

How To Grow Mushrooms - Learn About Growing Mushrooms

Not many people realize that it is straightforward to grow mushrooms yourself at home, instead opting to spend their money at their local supermarket on mushroom species cheaply imported from foreign countries where they are grown in bulk.

The shop variety does not have much of a shelf life, and the mushrooms don't really like to be packed in plastic so by learning to grow mushrooms at home not only are you going to have fresher longer-lasting mushrooms but they will also most likely taste stronger and more mushroomy as the shop varieties tend to have a more watered-down flavor.

Another advantage of growing mushrooms yourself is that you aren't limited to the variety displayed in the shops - which usually consists of button mushrooms, Shiitake, Oyster, and Portobello. Although Oyster mushrooms are seen to be the most accessible type of fungus to cultivate, you may wish to try and grow something that most shops won't ever sell. The Lions Mane mushroom is a little harder to grow and yet has a taste that is very similar to that of lobster, and it is costly to purchase from specialist retailers.

To be able to grow your own mushrooms first, you will need to decide on a variety. There are hundreds of edible mushrooms that can be grown either inside your house or outside, most growers settle for the oyster mushroom, to begin with, due to the simplicity of growing it (Oyster or Pleutorus Ostreateus has very vigorous growth and so is very likely to grow given the right conditions).

Once you have decided on a type of mushroom to grow, you will need to find the specific growing requirements, as all fungus have their different growing parameters. With the Oyster mushroom, you can use either a wood-based substrate (paper, cardboard, etc.), or you can grow it on straw. These are the most common substrates to use as they provide the best yields.

The next thing you will need is the mushroom spawn. It is easiest if you purchase your spawn from a shop - which is probably easiest done online as most garden centers only sell complete mushroom growing kits, whereas the spawn on its own is a little more specialist. Many websites sell spawn, and it will only cost you a few pounds for a bag that will provide you with lots of mushrooms (it is also far better value to grow your mushrooms than to purchase them from a store).

With the oyster mushrooms, you need to pasteurize the straw or paper-based product, which kills off many of the bacteria presents, giving the mushroom spawn a head-start when it comes to growing. You can do this by submerging the straw/paper in some hot water, keeping it at around 60 degrees C for about 1 hour. When this has done, drain the substrate and allow it to cool before loading it into a see-through plastic bag. Put a handful of straw/paper into the bag and then sprinkle spawn on top, and continue this until the container is full.

Tie the bag with a metal-tie and then pierce holes over the kit which will allow air to help the mycelium grow and will enable mushrooms to develop later, Leave it in a warm room for about two weeks until the bag completely colonizes (turns white, from the mycelium growing). An airing cupboard or boiler room is an ideal place).

When the bag is fully colonized, it will be ready to fruit - mushrooms should start appearing within a few days. To help it to the fruit, you need to move the bag to a more relaxed, damper area where humidity levels are about 90% or higher.

Oyster mushrooms like to be in quite cold conditions, so it is probably best to place them outside. They will start to form (pin) from the holes that were poked in the bag previously, due to the mushrooms liking the air provided.

When this happens, carefully cut the bag and peel it back a little, allowing the mushrooms the air and space required to grow to enormous sizes. When the Oyster mushrooms look the right size and just before the caps unfurl to release their spores, gently pull and twist them at their stems to harvest them. Cut the end part of the stem with a knife, and they will be ready to eat!

Types Of Mushrooms

With over 38,000 different varieties of mushrooms, anyone who is mad about mushrooms sure has plenty to choose from. Take a look at the distinct characteristics of some of the more popular varieties and a few handy tips for preparing your favorite type.

➢ Girolle Or Chanterelle Mushrooms

Trumpet-shaped and yellow-gold, Chanterelle mushrooms have a rich flavor that ranges from apricot to earthy. They are best eaten fresh, but are also available canned or dried.

➢ Enokitake Or Enoki Mushrooms

Native to Japan, Enokitake mushrooms have a sprout-like appearance with thin, long stems and small caps atop them. They are typically white, have a light, fruity taste, and are served raw in salads and soups.

➢ Crimini Mushrooms

Also called Italian Brown because of their dark brown color, Crimini Mushrooms are dense in texture and possess a rich flavor.

➢ Agaricus Mushrooms

More popularly known as button mushrooms or white mushrooms, these are the most commonly used variety. They are abundantly available fresh, frozen, or canned at any supermarket. Agaricus mushrooms have a mild flavor if eaten raw, but the feeling gets intensified when they are cooked.

➢ Porcino Mushrooms

Considered one of the finest among mushrooms, their high price is no deterrent to those who love the meaty texture and distinctive flavor of the Porcino mushroom. They are available in variable sizes and have a very characteristic shape.

➢ Shiitake Mushrooms

Cultivated initially only in Japan on natural oak logs, shiitake mushrooms are now available in larger grocery stores around the world. Large in size and black-brown in color, shiitake mushrooms have an earthy-rich flavor. Dried shiitakes have a more intense flavor and are often referred to as the fresh variety. They are commonly used in soups and stir-fried dishes.

> **Portobello Mushrooms**

Large, circular and flat, Portobello mushrooms can sometimes grow to the size of a regular-sized hamburger. They have a dense, chewy texture and are an excellent choice for roasting and grilling.

> **Pleurotus Mushrooms**

Pleurotus mushrooms, also known as oyster mushrooms because of the remarkable similarity in taste, are available in colors that range from off-white to different shades of brown. Their texture is more suitable for use in cooked dishes.

> **Morel Mushrooms**

Highly prized, with a price to match, these conical mushrooms with the porous surface are much sought after for flavoring stews and sauces with their intense earthy flavor. Morel mushrooms are small and dark brown and can also be used for stuffing.

Handy Hints And Tips For Preparing Mushrooms

> **Cooking With Mushrooms**

Remember, dried mushrooms have an incredibly concentrated flavor and should be considered more as seasonings rather than vegetables. Dried mushrooms need to be soaked in hot water for about 20 to 30 minutes, rinsed, chopped, and then added to soups, sauces, and stews.

Most mushrooms can be eaten raw as well as cooked. However, the stems of certain varieties included Portabella and Shiitake are often harsh and must be removed. They can be used as flavoring agents in individual dishes if need be.

Stuffed mushrooms are a much sought-after delicacy. For stuffing mushrooms, remove the stem, scrape out the gills and hollow out the fungus using a melon baller. You'll have plenty of space for any delicious stuffing.

➤ Mushrooms For Weight Loss: Nature's Secret Strategy In A Small Package

What has lots of nutrition, including protein, can be used in facials, brewed as tea, strewn on pasta, blended into smoothies, eaten alone, and used as medicine? You guessed it: MUSHROOMS. Yes, these versatile, beautiful living things that they are!

It is almost incredible to think that mushrooms, members of the Kingdom Fungi, can be so tasty, attractive looking, distinctive in smell, have 14 thousand species with several thousand edible, be so diversely nutritious, and have considerable medicinal properties.

The Institute of Medicine (IoM) is part of the National Academy of Sciences, which is non-governmental, commissioned with setting the Recommended Daily Allowances (RDA) that we all use as some measure of correctness, for nutritional values, even though we rarely question what criteria is tested and by whom.

According to one of the editors of these measurements, Professor Robert Reynolds, formerly of the University of Chicago, the system has many flaws because a small amount of money is allotted to studying this information.

Roberts says that only half of us fall into the "average" category for

Recommended Daily Allowances, and we have to eliminate the top 3% healthiest people to boot. The measure does not apply if we are sick, if we are overweight, if we are over 60, if we are stressed, if we take medication, if we smoke, if we eat refined and processed, unhealthy food that does not consist of 2,000 calories a day.

It makes me wonder how much it would cost to coordinate a redo of the RDA program using already existing research data with values for all the people the current system leaves out. Studies are massively expensive when done from scratch, so using secondary data would cut the costs.

The current Recommended Daily Allowances use sparse and outdated data, in many cases. But it is still useful.

The nutritional information that we see on packaging in the United States is accurate for healthy people who eat right, don't smoke, don't weigh too much, and don't get stressed out. And these values, themselves, are increased by about 25% as a buffer for the nutrition lost in cooking.

Nutritional Facts

The following nutrients make white mushrooms quite valuable and unique. The numbers in parentheses represent the percentage of daily needs in an average person.

- B Vitamins, aside from all their other benefits, are being looked at to reduce ADHD and slow Alzheimer's.
- Thiamin(e), B1, helps metabolize sugars and amino acids. (4%)
- Riboflavin, B2 metabolizes carbs into energy. (17%)

- Niacin, B3, increases the level of high-density lipids (HDL), the good cholesterol, in the blood. (13%)
- Pantothenic Acid, B5, turns carbs and fats into usable energy and assures healthy fats in cells. (10%)
- Pyridoxine, B6, balances sodium (Na) and potassium (K). (4%)
- Folic Acid, B9, is needed for DNA synthesis and repair and cell growth. (3%)
- Choline helps in cell membrane synthesis.
- Betaine regulates fluid movement across cell membranes (osmosis), assists in membrane work, and neurotransmission of acetylcholine.
- Omega-6 Fatty Acid is one of the two essential fatty acids required for cellular processes and must be in balance with Omega-3 Fatty Acids to keep inflammation down and cell function up. Both are essential and must be consumed. Therefore, get those walnuts and flax seeds to balance with Omega-3s.
- Copper assists with iron uptake and cell metabolism. It also protects our cardiovascular system. (11%)
- Phosphorus forms part of the structure of living molecules (DNA, RNA). (6%)

- Potassium helps the body process sodium. It is also essential in preventing muscles from contracting. You've heard someone recommend that you eat some dried apricots or a fig if you have leg cramps. They're high in potassium. (6%)
- Selenium helps with cell function.
- Vitamin D, which is produced by changing a sterol, ergosterol, into Vitamin D2, with ultraviolets present in the sun. (This is similar to the way we get vitamin D from the sun, except that we use 7-hydrocholesterol and synthesize Vitamin D3. (3%, which may be higher if the mushrooms were exposed to ultraviolet light)
- Mushrooms also contain 2% of the RDA for Vitamin C, Iron, Magnesium, Manganese, and Zinc.
- Mushrooms are about 15 calories for a cup, with 2 grams of protein and 2 grams of carbs, with only one counting since 1 of those grams is a total indigestible fiber that helps with digestion and does not turn to glucose before it gets to the colon.
-

➢ Other Health Benefits

- Research from Beckman Research Institute, Duarte, California, found that white mushrooms contain conjugated linoleic acid (CLA), which reduces high estrogen level risks like breast cancer. Other studies at the same institution showed that white truffles have a similarly beneficial effect on prostate cancer.
- Eating white mushrooms seems to block the production of inflammatory molecules, useful for reducing many types of inflammation, which is any immune compromise or sickness.
- Mushrooms, in general, are known for their antioxidant properties. L- ergothioneine, one of the antioxidants that are found in white mushrooms can be found in shiitake mushrooms with 5 X as much. So, different species are known to have various health accolades.
- Mushrooms have a low glycemic load (2), which means that they don't cause sugar levels to change rapidly. Their fibrous material is, in part, responsible for this.
- Inflammation Factor is low (-4), which suggests that they will not contribute to

inflaming the body. Mushrooms are also known for reducing it.
- Mushrooms are known to improve the immune system and cognitive function.
- The last and most important benefit for the WarriorsOfWeight.com Community is that THEY MAKE US FEEL FULL AND SATISFIED while we ingest very few calories and much nutritional value.

Cooked Vs. Raw

Many of us profess that natural foods have to have higher nutrients when fresh.

With mushrooms, according to research cited in Scientific American by Sushma

Subramanian, some veggies, including fungi, when boiled or cooked for a short time, increase antioxidants and other properties as the cell membranes break down and release the nutrients. Check out the recipe below, the way I've been eating mushrooms for years. And work the stems in somehow, even if they don't taste quite as creamy. Full of nutrition.

Toxicity Caution: Must Read

Mushrooms from the Agaricus Bisporus species, which is the species to which white mushrooms and many others we eat belong, reveal the presence of small amounts of hydrazine compound derivatives, agaratine, and gyromitrin. These substances are known carcinogens when delivered to mice in extremely high doses in short periods. There has never been a proven case that cancer has been caused by eating mushrooms.

Most researchers recommend that consideration be given to this fact and point out that cooking the mushrooms reduces the contents of the toxins.

Confirm that you do not feel allergic or headache from eating a small number of mushrooms before you eat a more significant portion. Sometimes, people can have adverse reactions to eating them. I had a slight headache from adding a mushroom supplement once. It was mild but lasted several days.

It should also be noted that, according to Joshua Rosenthal's nutrition program, mushrooms are in the top 15 foods with least residue from pesticides, although I do not know how he got this number. It's always better to eat organic food, which means it was not sprayed.

Action Steps

- Take a little time to read more on mushrooms. They are fascinating.
- Add a recipe of a small serving of mushrooms to your daily food several times this week, using caution as advised. Notice how the mushrooms fill you up. Or eat one mushroom a day.
- Make a special mushroom recipe that suits you well. Consider not adding sugar, fat, or salt. Give it to a friend or school mate.

The Mighty Mushroom

Every time my family ordered pizza when I was a kid, my dad would find some way to sneak mushrooms onto a corner of that pizza, possibly tucked under a layer of cheese per individual instruction to the pizza parlor. He loved them, but I was stubbornly convinced that a single mushroom would ruin the entire pizza.

Now that I'm older and have developed a more refined palate (ok fine, I still love Cinnamon Toast Crunch), I've come to understand how special mushrooms are in the world of cuisine. They provide an extraordinary variety of texture and flavors which seem to adapt to any dish. And, as a bit of icing on the cake, I've learned how mushrooms are lovely for your health. Consider this my ode to the fungus.

> ## Health Benefits

When it comes to health, edible mushrooms are a right up there with other super-foods green tea and broccoli. After all, the first antibiotics were extracted from fungi. Being 80-90% water, mushrooms are low in calories, while still being high in fiber. They are fat-free, cholesterol-free, and low in sodium (especially useful for those on a hypertensive diet). Here are some other reasons to sneak more mushrooms into your cooking:

Mushrooms are considered probiotic, meaning that they help the body to strengthen itself and ward off illness. Part of mushrooms' probiotic ability comes from their high percentage of the nutrient riboflavin. Mushrooms are a great source of potassium, a mineral that helps lower blood pressure and reduces the risk of stroke. A medium portabella mushroom has more potassium than a glass of orange juice or a banana.

Phytonutrients found in mushrooms have been at the center of anti-cancer research for decades. In many countries, medicinal mushrooms are used as an adjunct to other cancer treatments.

➢ White (Button)

White mushrooms range in color from white to light brown and come in many different sizes. The smaller varieties of white mushrooms are called button mushrooms and are easily the most popular mushroom in cooking, found in most grocery stores. Freshly picked white mushrooms have a mild or delicate flavor. As the caps darken, they develop a richer taste.

Recent studies have shown that white mushrooms can reduce the risk of breast and prostate cancer. Grilled Lemon Shrimp with Mushrooms

This healthy meal is perfectly seasoned with light lemon juice and garlic, grilled to perfection, and then stuffed into a pita. Carb-friendly and delicious!

Ingredients

- 8 oz. fresh white mushrooms
- 1 lb. jumbo shrimp, peeled and deveined
- Two medium-sized zucchini, sliced 1 inch thick (about 2 1/2 C.)
- One medium red onion cut in 8 wedges
- 1/4 C. olive oil
- 2 Tbs. fresh lemon juice
- 2 tsp. minced garlic
- 1 tsp. dried oregano leaves, crushed
- 1/2 tsp. salt
- 1/4 tsp. ground black pepper
- Four pitas, warmed
- Cucumber Yogurt Sauce:
- 1 C. plain low-fat yogurt
- 1 C. peeled, seeded and diced cucumber
- 1 Tbs. chopped fresh mint or parsley
- 1 tsp. minced garlic
- 1/2 tsp. salt

Directions

Preheat outdoor grill or broiler until hot. Leave small mushrooms whole; halve larger ones. In a large bowl, place mushrooms, shrimp, zucchini, and red onion. In a small bowl, combine olive oil, lemon juice, garlic, oregano, salt, and black pepper and pour over vegetables; toss until well-coated. Place vegetables and shrimp on a vegetable grilling rack or a rack in a broiler pan. Grill or broil no more than 6 inches from heat until vegetables and shrimp are just cooked, about 8 minutes, stirring often and occasionally brushing with remaining marinade. Serve on pitas with Cucumber Yogurt Sauce.

To make the sauce, in a small bowl, combine all ingredients and blend well.

- **Makes** about 1 1/2 cups.
- **Yield:** 4 servings
- **From:** St. Pete Times
- **Nutrition information per serving:** 308 calories, 25gm protein, 16gm fat

➢ Crimini/Cremini/Italian Brown

Crimini mushrooms are similar in appearance to white mushrooms but are a darker color, ranging from light tan to dark brown. They have a firmer texture and a stronger, earthier flavor than white mushrooms.

These make an excellent substitute for white mushrooms in any recipe and work exceptionally well with beef.

Crimini mushrooms are an excellent source of selenium, which is needed for the proper function of the antioxidant system in the body. Selenium works to prevent colon cancer, arthritis, and even asthma. Crimini mushrooms are also delicious as a source of zinc, a critical mineral for the immune system.

➢ Vegetarian Hobo Dinner

Cooked over hot coals, this meal is made with Boca "meat," mushrooms, carrots, and potatoes.

Ingredients

- Two carrots, sliced
- 6-8 new potatoes, quartered
- 1/2 onion, LG. chunks
- Two shallots, sliced
- 2-3 cloves garlic, LG. chunks
- 8-10 cremini mushrooms whole or halved
- 2-4 Tbs. olive oil
- 2 Tbs. unsalted butter, optional
- One pkg. frozen Boca ground "meat."
- salt & pepper, to taste
- season salt, dash

Directions

Mix all sliced vegetables in a bowl. Make two pockets with heavy-duty aluminum foil, doubled. Place a layer of plants on the bottom. Layer Boca ground "meat" next. Add a final layer of veggies. Pour 1-2 Tbs. of olive oil on each dinner, dot with butter, if using. Season with salt, pepper, and season salt. Fold foil to make an airtight seal. Cook on hot coals for fifteen minutes, flipping half-way through. Serve with ketchup, enjoy!

> **Portabella/Portobello**

Portabellas are a larger relative to white mushrooms, reaching a diameter of up to

6 inches. Portabellas take longer to spoil than white or crimini mushrooms.

Because of their longer growing cycle, they have a denser, meatier texture and flavor, making them delicious on sandwiches.

> **Portabella Pizza**

This delicious, crust-less pizza with cheese, tomatoes, and mushrooms is part of a low-carb diet.

Ingredients

- 1 to 2 tsp. extra-virgin olive oil
- One clove garlic, peeled, minced
- 6 oz. portabella mushroom caps (about 4), cleaned, stems removed
- Pinch of salt and freshly ground black pepper to taste
- 12 oz. shredded or sliced mozzarella cheese ten fresh basil leaves
- Two fresh tomatoes, sliced, roasted, grilled or broiled Oregano leaves, optional

Directions

Preheat the oven to 450 degrees. In a small bowl, combine the oil and garlic. Rub the mushroom caps on all sides with the oil mixture. Place the caps, top side down, in a circle on an oiled baking sheet. Season with salt and pepper. Arrange the cheese, basil, and tomato slices alternately in a circle on top of the mushrooms. Sprinkle with oregano, if you like. Bake until the cheese melts, about 3 to 5 minutes. Remove from the oven and serve.

➤ Shitake (Oak/Chinese/Black Forest)

Shitake mushroom caps have a rich, woodsy flavor and soft, spongy texture. They range in color from tan to dark brown with broad, umbrella-shaped caps. Shitake mushrooms can last up to 14 days, and then discarded items can be used to flavor soup stocks. Used for centuries in East Asia to fight colds and flues, shitake mushrooms have been shown to help stimulate the immune system, fight infection, and ward off tumors. Shitake also treats nutritional deficiencies and liver ailments.

Ingredients

- 4 C. water
- 1-2 C. chopped organic vegetables (see note)
- 1 1/2 Tbs. of dark organic barley
- Miso Firm tofu, diced into 1/2 inch cubes
- 3-inch piece dried wakame seaweed (found in most health food stores)
- Two shitake mushrooms, organic, dried (Can pre-soak per package instructions.)

Directions

Boil the water in a small pot. Add chopped vegetables and mushrooms to boiling water. Lower heat, cover, and simmer until vegetables are tender (about 8-15 minutes depending on the plants used). After plants have simmered for about 5 minutes, place a 1/4 C. hot vegetable broth from the pot in a separate bowl. Add miso to a bowl and mix until miso becomes a wet paste.

Add tofu to a bowl of miso mixture and set the pan aside until vegetables are tender. Tear seaweed into small pieces and add to the pot. When vegetables are tender, add the miso mixture from the bowl to the pot. Let stand for 3 – 4 minutes. Don't heat miso on high heat, as it will kill the living microorganisms that aid in digestion and healing. Remember making this healing soup is intuitive. You can try more or less miso and different vegetable combinations. Honor your body's wisdom as you experiment with this miso soup recipe. Enjoy this healing soup.

- **Chanterelles**

Many varieties of chanterelles are delicious in cooking, one of the most identifiable being the yellow chanterelle (pictured). Chanterelles have a delicate flavor and a more elegant texture, making them perfect for egg dishes and as a topping on pizza. The bioluminescent Jack-O-Lantern chanterelle is extremely poisonous to humans but not fatal.

➢ Snow Peas and Wild Mushrooms with Ginger

Excellent side dish, especially if you can find a variety of wild mushrooms.

Ingredients

- 1/2 tsp. canola or sesame oil
- 3 C. mushrooms, mixed (shiitake, chanterelles, oyster, etc.), sliced
- 2 Tbs. ginger root, peeled and sliced into 1/2" match sticks
- 1/2 tsp. ginger powder
- 1/2 tsp. cardamom, ground
- 1 tsp. cornstarch
- 2 tsp. rice vinegar
- 1/2 tsp. soy sauce, low-sodium
- 3 C. snow peas, fresh or frozen
- 15 oz. canned baby corn

Directions

Heat a sauté pan over medium-high heat and add oil. Sauté mushrooms, ginger root, ginger powder, and cardamom for 2 minutes, stirring occasionally. Dissolve cornstarch in vinegar and soy sauce. Add cornstarch mixture, snow peas, and baby.

Heat 2-3 minutes. Remove from heat and serve. Don't over-cook the vegetables.

➢ Porcini

Often considered one of the most beautiful mushrooms for cooking, porcini mushrooms are thick, meaty, and versatile. Italian cooks often season the mushroom with a woodsy variety of thyme called nipetella. Because of the heartiness of the mushroom, porcini do very well when dried (pictured).

➢ Noodles with Wild Mushrooms

This is an excellent dish for summertime when you can get a variety of wild mushrooms at your local farmer's market.

Ingredients

- 1 lb. of noodles (fresh or frozen)
- 12 oz. mushrooms (Portobello, shiitake or porcini), diced
- 8 Tbs. butter
- 2 Tbs. chicken broth
- 2 Tbs. beef broth
- 2 Tbs. parsley, chopped
- Salt and pepper to taste

Directions

Melt the butter in a skillet. Add the mushrooms and sauté until just soft. Add the broths and parsley. Cook the noodles in a separate pot. Drain and toss with the mushroom sauce. Reheat if necessary. Check the seasonings. Serve warm.

➤ Oyster

Oyster mushrooms have broad, fluted caps, and are described as graceful by many. Often growing on the sides of trees, these mushrooms are most commonly white but can also feature more exciting colors in the wild like pink or yellow.

They have a mild flavor and the most velvety texture of any mushroom.

Oyster mushrooms have a protein quality almost equal to animal-derived protein without the fat. These mushrooms have also been shown to work against cholesterol.

➤ Mushroom And Chestnut Soup With Roasted Fennel

This elaborate soup uses three different kinds of mushrooms, and the fennel gives it some unique flavoring as well. A very sophisticated soup.

Ingredients

- Three fennel bulbs stalks cut off, in half
- 3 Tbs. olive oil
- 1/4 tsp. sea salt
- 1/8 tsp. black pepper
- 3 Tbs. butter
- One diced white onion
- Three cloves garlic, minced
- 1 1/2 tsp. chopped thyme
- 1 7 oz. jar whole peeled chestnuts
- 2 10 oz. Pkg. white button mushrooms, sliced
- 8 oz. oyster mushrooms, sliced
- 8 oz. chanterelle mushrooms, sliced
- 1 tsp. sea salt
- 1/4 tsp. black pepper

- Two qt. chicken broth
- 1/3 C. heavy cream

Directions

Heat oven to 400 degrees. Cut fennel bulbs, stalks cut off, in half, place cut side up on the baking sheet. Drizzle with olive oil, sprinkle with 1/4 tsp. Sea salt and 1/8 tsp. Black pepper. Roast until tender, 30 minutes. In 8 quart pot, melt butter, add onion, and garlic, cook 5 minutes. Add thyme, chestnuts, white button mushrooms, oyster mushrooms, chanterelle mushrooms, sea salt, and black pepper. Cook until mushrooms are wilted, 15 minutes.

Remove 1 C. chestnuts and mushrooms, coarsely chop and reserve for garnish.

Pour 2 quarts chicken broth into the pot, bring to boiling. Reduce heat, simmer 15 minutes. Remove from heat. You are using a blender, puree soup until smooth.

Stir in 1/3 C. heavy cream, keep warm.

To serve: Remove and discard tight core from fennel halves, chop fennel. Spoon soup into bowls. Drizzle each serving with olive oil and garnish with chopped fennel and mushroom mixture.

Morel

Morel mushrooms have spongy caps resembling honeycombs and short, thick stems. Morals have a rich, nutty taste and a robust and woodsy fragrance. One variety of morel called the False Morel (pictured), is deadly poisonous when eaten raw but considered a delicacy in some parts of the world after cooking.

➤ **Red Wine-Braised Rabbit With Wild Mushrooms**

Ingredients

- Two fresh rabbits, cut into serving pieces
- Marinade:
- One medium onion, sliced
- 1 C. red wine, such as Syrah or Cotes-du-Rhone
- 1 Tbs. olive oil
- Three garlic cloves, crushed
- Four juniper berries, toasted and coarsely ground
- Two rosemary sprigs, coarsely chopped
- Two thyme sprigs
- Kosher salt and freshly ground black pepper
- Braise:
- 1 Tbs. olive oil, or as needed
- One heaping C. diced carrots

- Two garlic cloves, finely minced
- 3 or 4 slices dried porcini mushrooms, rinsed and soaked for 30 minutes in 1/2 C. hot chicken broth or water
- 2 C. red wine, such as Syrah or Cotes-du-Rhone
- 1/2 C. port
- 2 C. chicken broth or canned low-sodium food
- Four thyme sprigs
- Six flat-leaf parsley sprigs
- One leek top
- Two bay leaves
- 3 Tbs. unsalted butter
- 1/4 lb. fresh porcini, morel, chanterelle, or cremini mushrooms, trimmed
- Spoonbread
- Chopped flat-leaf parsley, basil, or thyme for garnish

Directions

Place the rabbit in a shallow ceramic or another non-reactive dish. In a small bowl, combine all the marinade ingredients. Pour the marinade over the rabbit, turning to coat, cover, and marinate in the refrigerator for 6-24 hours. Remove the rabbit from the marinade and set aside. Strain the marinade into a bowl, reserving the vegetables; place the liquid and vegetables alone.

Preheat the oven to 325 degrees. Pat, the rabbit, pieces dry and season them with salt and pepper. In a deep heavy ovenproof skillet or a Dutch oven, heat the oil over medium-high heat. Add the rabbit pieces, in batches, being careful not to crowd the skillet, and sear, turning once until they are golden brown, 10-12 minutes; carefully monitor the heat so that the oil does not burn, adding more oil between batches if necessary. Transfer the rabbit to a rack set over a baking sheet.

Tie together thyme, parsley, look top and bay leaves to make a bouquet garni and set aside. Add the carrots, garlic, and reserved vegetables from the marinade to the skillet and cook over medium heat until softened about 10 minutes. Add the dried mushrooms and their liquid, the wine and port, and the reserved liquid from the marinade. Bring to a simmer and skim off any foam. Add the broth, bouquet garni, and the rabbit and bring back to a simmer. Cover tightly with the lid or aluminum foil and place it in the oven — Cook for 15 minutes.

Remove the loin pieces and set aside. Continue to braise the remaining rabbit for 30-40 minutes, or until tender. Transfer the rabbit pieces to a rack set over a platter. Place the skillet half on and half of a burner (this will make skimming off the fat easier) and bring to a boil over medium-high heat. Boil to reduce by half, skimming frequently.

Then strain the liquid into a saucepan, reserving the vegetables. Discard the bouquet garni and puree the plants through a food mill. Add the pureed vegetables to the skillet, bring to a simmer, and reduce until the sauce is thick enough to coat a spoon. Adjust the seasoning.

Meanwhile, in a small sauté pan, melt the butter over medium-high heat. Add the fresh mushrooms and sauté until some of their juices are released, but they are still firm 3-4 minutes.

Remove From The Heat. Just Before Serving, Re-Warm The Rabbit In The Sauce.

Place a helping of spoonbread in the center of each plate and surround it with the rabbit and mushrooms. Spoon the sauce over the rabbit and garnish with the chopped herbs.

Mushroom Varieties and Their Uses

Mushrooms are a unique source of food and come in infinite varieties. They are a type of living organism that has no roots, leaves, flowers, or seeds. Mushrooms are fungi, and in many countries, that is what they are called. There are countless varieties of Mushrooms that are edible, and there are probably just as many or more that are not edible.

The mushrooms that are not edible can be poisonous and can cause severe illness or, worse, death. For that reason, wild mushrooms should not be picked by anyone other than a trained mycologist. Mushrooms can be purchased dried, canned, or fresh. For a long time, even though there are over 590 species of Mushrooms found growing in California, the only Mushrooms readily available in the United States for consumption were Brown Mushrooms and White Mushrooms.

Photos of some of the California Mushrooms can be found on Myko Web, a site that specializes in California Mushrooms. Some mushrooms are so amazingly beautiful that it is hard to believe that they can be poisonous. With the increasing population growth from Asia and the Middle East and the rise of the

Television Food Shows, our food selections have significantly increased.

Today, you can walk into almost any Supermarket and find at least half a dozen varieties of mushrooms readily available. Some of the varieties that you can purchase are Crimini, which is small brown mushrooms, Portobello, which are a larger version of the Crimini, White Mushrooms, Shitake or Wood Mushrooms, Oyster, Enoki, Chanterelles, and Truffles. Mushrooms can be cooked whole, quartered, sliced, or chopped.

The Crimini are excellent in stews, sauteed or stir-fried with other vegetables. They are also good served with steak or different types of meat. The Portobello, which is essentially a fully grown Crimini, is great for Mushrooms burgers.

Remove the stems, marinate them in Italian Salad Dressing and then grill them on a stovetop grill and serve on Hamburger Buns topped with Provolone Cheese and Lettuce and Tomato.

You will get the same satisfaction from this Burger as one with beef, and it is much healthier for you. The Portobello can also be stuffed with either a crab stuffing, a breadcrumb, and chopped mushroom stem stuffing or rice stuffing. They are large enough so that one stuffed mushroom can serve as an entre for dinner served along with a salad.

The White Mushrooms can be used in pretty much the same way as the Crimini. The larger of the White Mushrooms are great stuffed for appetizers. They can be served at the table or passed around as finger food before dinner starts or be part of a great appetizer party. The Shitake, which is sometimes called tree mushrooms or forest mushrooms, is native to East Asia. They have a unique taste, which can best be described as a blend of filet mignon and lobster.

Unlike the stems of the Brown and White Mushrooms, the stem portion of the

Shitake is too tough and woody to eat, so they should be removed before preparing. Oyster mushrooms grow in clumps and do have the shape of an oyster, and they have a chewy texture. It is not entirely sure how Oyster Mushrooms got their name. Some say it is from the flavor, and others say that it is from their shape. They were first cultivated in Germany during World War I as a subsistence measure.

They are now cultivated worldwide and are especially favored in Asian countries. Enoki Mushrooms were initially a delicacy in Ancient Japan and were limited to a farming region in the northern part of the country where they grew wild. The Enoki is named for the tree on which it grows in the wild. This is a Japanese Hackberry Tree or Enoki, which is its Japanese name.

It has only been in the last few decades that a technique was developed to cultivate them and grow them in other areas. The Enoki is popular in Japan, China, and Korea. These miniature, slender mushrooms have a crisp texture and create a lovely, fragile, delicate visual effect.

The Chanterelle Mushroom is orange or yellow, meaty, and funnel-shaped. It has a fruity smell, somewhat like apricots, and a mildly peppery taste and is considered an excellent edible mushroom. The Chanterelle is common in

Northern Europe, parts of North American, and Mexico and can also be found in Asia and Africa. Many popular methods of cooking chanterelles include them in sauces, souffle, cream sauces, and soups. They are not typically eaten raw, as their rich and complex flavor is best released when cooked.

When using truffles in cooking, usually only a minimal amount is needed. The truffle is shaved into thin slices and used as a seasoning for a dish. Truffle Oil can be purchased and used for culinary purposes also. The advantage of using the oil is that it will last longer than a fresh truffle will, and the flavor won't be quite as strong.

Mushrooms are nutritious food in that they are low in calories, contain no fat, and contain significant amounts of protein and Vitamin C. Therefore, they are great for dieters, but aside from that, they add a marvelous flavor and texture to food. They can be cooked by themselves as a side dish; they can be used as entrees or to add flavor and texture to other dishes. To prepare mushrooms for cooking, follow basic rules for retaining maximum character.

- **Do Not Wash:** Wipe with damp paper toweling.
- **Do Not Peel:** The skin is very flavorful. If there are any bad spots, cut off only that spot and no more.
- **Do Not Soak:** If you do, they will become waterlogged, losing vitamins, as well as flavor. Remember that mushrooms will add flavor and texture to your food and can be a tasty dish by themselves.

With the addition of a little garlic, fresh basil, and olive oil, you can create dishes that will be enjoyed by all.

Conclusion

There are so many benefits to mushrooms. They can be used in so many ways: as antioxidants, antivirals, an anti-migraine remedy, for anti-cancer treatment, as anti-psychotics, for daily nutrition, improved memory, and beauty care. The wealth of vitamins, minerals, essential fatty acids, and overall healthy properties make it a superfood. After seeing all of what it does, consider telling everyone you're adding it to your day.

As I said at the beginning, this book is not meant to be exhaustive, but was written to serve as a general pointer in the right direction for anyone who is looking for such information, and also a reference for the future. This book is an excellent summary of where, in our economic, social, and ecological theories and practices, we have gone wrong and how we might start to learn to notice what is happening and then humbly take our place in the world.

What needs to be crystal clear is that different people will have different experiences, even if they consume the same psychedelics in the same setting.

These experiences are unique every time, and in a way, every psychedelic experience is like the first time –

because you never know what you are going to experience.

Everyone experiences what they need to experience, even if it may be unpleasant. With all precautions taken, at the end of the day, how you are as a person will significantly determine your magic mushroom experience, not vice- versa. This means that magic mushrooms are merely tools (albeit mighty ones) for self-exploration, but other than that, it is you who is creating the experience in itself.

Make sure that you consume these powerful medicines with someone you trust, ideally someone who knows how to facilitate such an experience and be careful with the dosage. It is better to start with a smaller dose than with a bigger one. The set and setting in which you consume magic mushrooms is of crucial importance and can make the difference between an empowering life-changing experience and a traumatic experience.

I Hope You Enjoy Reading This Fantastic Guide!

Made in the USA
Monee, IL
12 April 2020